Reservoir Sediment Management

Reservoir Sediment Management

Şahnaz Tiğrek

Civil Engineering Department, Middle East Technical University,
Ankara, Turkey

Tuçe Aras

EKON Industry Construction & Trade Inc., Ankara, Turkey

CRC Press

Taylor & Francis Group
Boca Raton London New York

CRC Press is an imprint of the
Taylor & Francis Group, an **informa** business

A BALKEMA BOOK

Cover illustration: Anatolia – Ataturk Dam on the Euphrates River, Turkey

CRC Press
Taylor & Francis Group
6000 Broken Sound Parkway NW, Suite 300
Boca Raton, FL 33487-2742

First issued in paperback 2018

CRC Press/Balkema is an imprint of the Taylor & Francis Group,
an informa business

ISBN-13: 978-0-415-60339-3 (hbk)
ISBN-13: 978-1-138-37255-9 (pbk)

Typeset by MPS Limited, a Macmillan Company, Chennai, India

Library of Congress Cataloging-in-Publication Data

Tiğrek, Şahnaz.
    Reservoir sediment management / Şahnaz Tiğrek, Tuçe Aras. — 1st ed.
       p. cm.
    Includes bibliographical references and index.
    ISBN 978-0-415-60339-3 (hard cover : alk. paper)
    1. Reservoir sedimentation.   2. Sediment transport.
I. Aras, Tuce. II. Title.

    TD396.T54   2011
    627'.86—dc23
                                                                            2011027243

Published by: CRC Press/Balkema
                P.O. Box 447, 2300 AK Leiden, The Netherlands
                e-mail: Pub.NL@taylorandfrancis.com
                www.crcpress.com – www.taylorandfrancis.co.uk – www.balkema.nl

Visit the Taylor & Francis Web site at
http://www.taylorandfrancis.com

and the CRC Press Web site at
http://www.crcpress.com

Contents

List of Figures

List of Tables

List of Symbols

A	flow area
A_f	cross sectional area of valley scoured out by flushing (m^2)
A_r	cross sectional area of reservoir in reach immediately upstream from the dam (m^2)
C_o	reservoir capacity (m^3)
C_u	uniformity ratio
C_z	Chezy coefficient
d	diameter of the sediment particle
d_{av}	average grain diameter of the sediment particles
d_{50}	median particle size of sediments
DDR	drawdown ratio
DX	distance between the two adjacent stations.
El_f	water surface elevation at the dam during flushing (m)
El_{min}	minimum bed elevation (m)
El_{max}	top water level elevation (m)
Fr	Froud number
FWR	flushing width ratio
g	gravitational acceleration
h	flow depth
H	water depth at the dam section
h_f	a height defined in Figure 2.8 (m)
h_l	a height defined in Figure 2.8 (m)
h_m	a height defined in Figure 2.8 (m)
h_n	normal depth
L	reservoir length (m)
LTCR	long term capacity ratio
MAR	mean annual river runoff
M_f	sediment mass flushed annually from the reservoir (t)
M_{dep}	sediment mass deposited annually in the reservoir (t)
M_{in}	mean annual sediment inflow (m^3)
MSY	mean annual sediment yield
n	Manning's roughness coefficient
ND	number of reaches
NS	number of stations
p	porosity

P	wetted perimeter
q	water discharge per unit width
q_s	sediment discharge
q_t	total-load discharge per unit width
Q_f	representative discharge passing through reservoir during flushing (m³/s)
Q_s	sediment load during flushing (t/s)
R	hydraulic radius
Re_*	critical grain Reynolds number
Re	Reynolds number
S	longitudinal river bed slope
SBR	sediment balance ratio
SBR_d	sediment balance ratio calculated for full drawdown
SEPS	sediment evacuation pipeline system
S_e	slope of energy grade line
SS_s	representative side slope for the deposits exposed during flushing
SS_{res}	representative side slope of reservoir
S_t	remaining reservoir capacity after year t
S_w	slope of the water surface
t	time
TE	trapping efficiency
T_f	duration of flushing (days)
TWR	top width ratio
U	flow velocity
u_*	shear velocity
ν	kinematic viscosity of water
V	average flow velocity
V_{in}	mean annual inflow volume (m³)
w_s	fall velocity (m/s)
W	the representative width of flow for flushing conditions (m)
W_{bf}	scoured valley bottom width for complete drawdown (m)
W_{bot}	representative bottom width for the reservoir (m)
W_f	width of flow at the bed of flushing channel (m)
W_{res}	representative reservoir width in the reach upstream from the dam at the flushing water surface elevation (m)
W_t	reservoir width at the top water level (m)
W_{td}	width at top water level of the scoured valley, when drawdown is complete (m)
W_{tf}	scoured valley width at top water level for complete drawdown (m)
x	longitudinal distance
γ	specific weight of water
z	vertical distance
σ_g	geometric standard deviation
θ_*	tractive force coefficient in Shields' diagram
τ	bed shear stress
τ_c	critical bed shear stress
θ_{*c}	dimensionless critical shear stress parameter of Shields'

Abbreviations

ASKİ	General Directorate of Ankara Water and Canalization Administration
BO	Build – Operate
BOT	Build – Operate – Transfer
CAT	Caterpillar
DSI	State Hydraulic Works
EPDK	Energy Market Regularity Authority
EÜAŞ	Electricity Generation Co. Inc
GIS	Geographical Information System
HSRS	Hydrosuction Sediment Removal System
ICOLD	International Commission of Large Dams
IRTCES	International Research and Training Centre on Erosion and Sedimentation
N/A	Not Applicable
NPV	Net Present Value
O&M	Operation and Maintenance
ppm	Parts Per Million
RESCON	Reservoir Conservation
RMB	Renminbi (Chinese Yuan)
SEPS	Sediment Evacuation Pipeline System
TEDAŞ	Turkey Electricity Distribution Co. Inc.
TEİAŞ	Turkey Electricity Transmission Company
TETAŞ	Turkey Electricity Trade Co. Inc.

Foreword

Sustainable development of river basins has been subject of debate for more than a decade, due to the increasing world population, polluted water resources and climate change. After 150 years of dam construction, only a small number of water basins have not yet been explored, and tools are needed to maintain and manage the basins that have already been developed. With many of the World's rivers being occupied by more than one dam, the conservation and sustainable management of existing reservoirs seems more effective than the construction of new dams as well. Still, siltation is the biggest threat to a dam, and good and effective siltation management of a reservoir is considered an indication of the dam operation's level of sustainability. Therefore, the control of sediment accumulation in dam reservoirs is essential to realize effective basin water storage capacity

In this book, practical tools are offered to control reservoir sedimentation that can be applied by professionals in planning, designing and operating dam basins. Although theoretical knowledge of the complex subject of sedimentation is essential in understanding the phenomenon thoroughly, this publication covers the hands-on dealing with sedimentation, and how to control it, in and around water structures.

After a brief review of sedimentation problems worldwide, empiricial and one-dimensional models are provided to calculate the life of a reservoir. These models can be applied rather easily and will also be useful in the planning of a dam. Then, the management of reservoir sediment is extensively described and explained. With the help of the RESCON open access software, resulting from a research project funded by the World Bank, the reader can examine and design sediment removal strategies such as flushing, hydrosuction, sediment removal, dredging and trucking for the operation, maintenance and design of several types of dams.

The authors,
August 2011

Acknowledgment

The authors' sincere appreciation goes to the World Bank for giving usage permission for RESCON. We also would like to thank Dr. Mustafa S. Altınakar (University of Mississippi) for his sincere support. Furthermore, we kindly acknowledge our colleagues, Şahnur Yılmaz Altun (M.Sc.), Oğuzkağan Çetinkaya (M.Sc.), Özge Göbelez (M.Sc.) and Başar Pulcuoğlu (M.Sc.) for their valuable contributions.

## Chapter 1

Conceptual framework of reservoir sedimentation

1.1 WORLD RESERVOIR SEDIMENTATION

Water is a vital component for the continuity of life for human beings. In history civilizations have settled at riversides in order to get their basic needs in daily life. However, the scarcity of water due to population increase, climate change, drought etc., causes a need to store fresh water. As a result, dams were constructed to store the water. Dams have been used for flood prevention too. Moreover, they have been used for power generation purposes for the last 100 years. As the dams are getting older some environmental damages have been observed. One of the reasons for this is the trapping of sedimentation on the upstream of the dam body.

Reservoir sedimentation is a very serious problem for many countries, especially in semi-arid regions. There are approximately 45,000 dams in the world and they lose their reservoir capacity by 0.5–1.0% every year (Mahmood, 1987). Moreover, in arid climate conditions the capacity lost is as high as 6,000–8,000 m^3/km^2/year (White, 2000).

The deposited sediment in reservoirs causes not only reservoir capacity loss but also downstream and upstream negative influences. In addition, the operation performance of the dam decreases significantly, such that the benefits gained from hydroelectric power generation, irrigation, water supply and flood control decreases a considerable amount because of the reservoir sedimentation. Moreover, the downstream and upstream faces of the dam are indirectly affected. In the downstream sediment loss causes degradation in channel and changes in aquatic life. That is why normal river conditions change the sediment-free water conditions due to sedimentation deposition in the upstream. On the other hand, increase of local ground water table, channel flood capacity reduction, decrease of bridge navigational clearance, and water diversions and withdrawals influence are negative effects of sedimentation in the upstream of a reservoir (Fan, 1985).

The International Commission on Large Dams (ICOLD) classifies a dam as a large dam if the height is higher than 15 m from the foundation or the volume of the reservoir is equal to, or more than, 3,000,000 m^3. There are 45,000 large dams around the world and China has the most: 22,000 in total. As far as the Mediterranean Area is concerned: Spain takes first place with 1,196 dams; Turkey follows Spain with 625 dams; and France, with 566 dams, is the third in ranking. Table 1.1 shows the storage, power and sedimentation of dams according to region worldwide.

Table 1.1 Worldwide Storage, Power and Sedimentation (RESCON Manual Volume I, 2003, after White, 2001)

Region	Number of large dams	Storage (km³)	Total Power (GW)	Hydropower Production in 1995 (TWh/yr)	Annual loss due to sedimentation (% of residual storage)
Worldwide	45,571	6,325	675	2,643	0.5–1
Europe	5,497	1,083	170	552	0.17–0.2
North America	7,205	1,845	140	658	0.2
South and Central America	1,498	1,039	120	575	0.1
North Africa	280	188	4.5	14	0.08–1.5
Sub Saharan Africa	966	575	16	48	0.23
Middle East	895	224	14.5	57	1.5
Asia (excluding China)	7,230	861	145	534	0.3–1.0
China	22,000	510	65	205	2.3

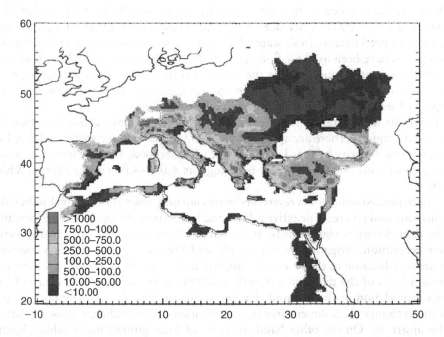

Figure 1.1 Increase of water storage capacity of the riparian countries of the Mediterranean Sea (Ludwig, 2007)

Ludwig (2007) shows water storage capacity distribution in Mediterranean Region during the second half of the 20th century in Figure 1.1. In addition, Poulus and Collins (2002) examined 69 rivers out of 169 rivers of the Mediterranean drainage basin, which incorporates more than 160 rivers with catchments greater than 200 km². They concluded that sediment supply is reduced by approximately 50% because of the construction of hundreds of dams around the Mediterranean Sea. As a matter of fact,

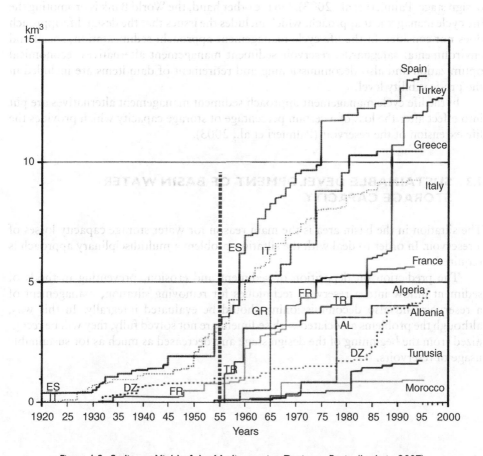

Figure 1.2 Sediment Yield of the Mediterranian Drainage Basin (Ludwig, 2007)

this reduction is mainly responsible for the loss of deltaic land. Further, as shown in Figure 1.2, a dramatic increase in the number of dams can be seen during the second half of the 20th century.

1.2 CHANGING PARADIGM FROM DESIGN LIFE APPROACH TO A LIFE CYCLE APPROACH

Generally, the design life approach is adopted by design engineers – the sedimentation accumulation is only used for economic life calculation. At the design stage, the benefit–cost calculation is performed using a specified time. For instance, Dams in Turkey have been designed for a 50 year economic life period, whereas it can change depending on the local circumstances. This design life is defined as the feasible operation time period of the project. The environmental and social issues of the project are included only at the initial stage. In addition any changes during this design life are not included in the design stage (Palmieri et al., 2003). On the other hand, the World Bank is promoting the life cycle management approach, which includes the issues that the design life approach does not consider. In the life cycle management approach; sedimentation, social and environmental safeguards, reservoir sediment management alternatives, economical optimization and also decommissioning and retirement of dam items are included in the pre-feasibility level.

In the life cycle management approach sediment management alternatives are put into effect after the loss of a certain percentage of storage capacity which provides the life extension of the reservoir (Palmieri et al., 2003).

1.3 SUSTAINABLE DEVELOPMENT OF BASIN WATER STORAGE CAPACITY

The siltation in the basin area is the main reason for water storage capacity losses of a reservoir. In order to deal with the siltation problem a multidisciplinary approach is required.

The prediction on deposition of sediment and erosion, preventing methods of sediment inflow into a reservoir, techniques for removing siltation, management of a reservoir and also decommissioning should be evaluated integrally. In this way, although the problems associated with sediment are not solved fully, they will be recognized from the beginning of the design stage and decreased as much as for sustainable usage of reservoirs.

## Chapter 2

Theoretical aspects of sediment transport

Deposition of sediment particles creates major problems in the operation of water resource systems, due to reservoirs' essential role in acting as sediment traps, interrupting fluvial sediment transport by producing an almost motionless pool into which flowing solids are deposited. Sediment accumulation in the active volume will gradually reduce the useful storage of the reservoir and raise the backwater profile elevations at the upstream. As a consequence, the reservoir lake area will increase but the water stored in the reservoir will not be sufficient for the intended purposes. In addition, continuous piling up of the sediment will deteriorate the quality of the water and the end result will be non-functioning intakes.

When a watercourse reaches a deeper water body such as a sea, lake or a reservoir of a dam, it starts to fail to keep its sediment carrying capacity due to the decrease in velocity. The coarser particles deposit at the entrance where the watercourse flows into to the water body forming, in general, a triangular-shaped depositional formation called a delta. Relatively finer sediment particles are carried further and the finest particles reach the dam structure with density currents. The dam body consists of intake structures at different levels according to their functions such as power generation and/or water supply as seen in Figure 2.1. An outlet at the upper layer (outlet C in Figure 2.1) may act as an intake structure to a turbine, which needs clear water, another outlet (B) may divert water for domestic or irrigation purposes. Thus, the initiation of a muddy lake is an undesired situation at each case. Further, sediment accumulation in the dead volume of a dam after a certain time may prevent the operation of the bottom outlets (outlet A in Figure 2.1) which are needed to discharge excess water. In some situations, dams are abandoned due to sedimentation, even if they are not completely silted.

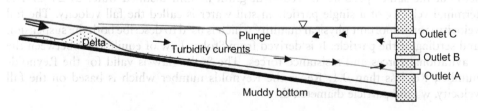

Figure 2.1 Sedimentation Process within a Reservoir

Therefore, estimation of sediment deposition rate, and the amount and location in reservoirs, is an important task in the design of hydraulic structures since sedimentation causes economical problems. However, prediction of the sediment accumulation is a complex task since it involves lots of variables, such as the texture and size of the sediment particles, reservoir operation schedule, seasonal variations in river flow, size and shape of the reservoir, etc. (Chen et al., 1978).

2.1   SEDIMENT PROPERTIES

Sediment is the end-product of erosion of the land surface by the combination of gravity effect with one or more than one climatic condition, such as rain, snow melt, wind and ice. This product can have some common characteristics depending on the surface texture and the mechanism of occurrence. Sediment properties of individual particles (size, shape, fall velocity and chemical composition) and bulk properties of particles (particle size distribution, porosity, specific weight and angle of repose) are used to classify sediment.

Sediment grain size can vary over several orders of magnitude. The most common size classification system which is recommended by American Geophysical Union (AGU) is given in Tables 2.1a and 2.1b (Lane, 1947).

Each sediment grain also has a range of dimensions that could be taken as the diameter. Different measurement techniques are used for different types of sediment; from manual methods for coarse particles, to sieving for gravels and sands, to more intense laboratory methods for fine-grained material. In brief they can be reduced to three dimensions of the sediment particle in the direction of three mutually orthogonal axes. They are ordered as the largest (a), the medium (b) and the shortest (c) dimension (Yang 1996 and Harris 2003) Arithmetic mean of a, b, and c is called a tri-axial diameter. The sieve diameter represents the diameter of the smallest circle that encompasses one dimension of the grain. It is the scale of the sieve mesh that would trap the sediment. It would be close to c and b. The nominal diameter represents the diameter of the sphere that would take up the same volume as the sediment grain. It would be somewhere in between a and c.

Since sediment particles are more than often accepted as spherical although there are never exactly spherical. Shape factor, SF definition based on three axes $SF = c/\sqrt{ab}$ is commonly used. If the particle is complete sphere, the shape factor will be equal to one. Shape factor of a natural sand is 0.7 and gravel is 0.9 (Depeweg and Mendez, 2007).

The standard fall diameter represents the diameter of the quartz sphere that would settle at the same speed as the sediment grain in still, distilled water at 24°C. The terminal velocity of a single particle in still water is called the fall velocity. The fall velocity is an important physical quantity which is used to describe both the suspension and settling of the particle. It is derived from Stokes Law of equilibrium between the gravitational forces and resistance forces. The Stokes law is valid for the Reynolds number, Re_f less than 0.1. Re_f is the Reynolds number which is based on the fall velocity, w_s and particle diameter, d.

$$Re_f = w_s d/v \qquad (2.1)$$

Table 2.1a Sediment Grade Size (Lane, 1947)
greater than 2 mm

Millimetres	Class
4000–2000	Very large boulders
2000–1000	Large boulders
1000–500	Medium boulders
500–250	Small boulders
250–130	Large cobbles
130–64	Small cobbles
64–32	Very coarse gravel
32–16	Coarse gravel
16–8	Medium gravel
8–4	Fine gravel
4–2	Very fine gravel

Table 2.1b Sediment Grade Size (Lane, 1947)
less than 2 mm

Millimetres	Class
2–1	Very coarse sand
1–0.5	Coarse sand
0.5–0.25	Medium sand
0.25–0.125	Fine sand
0.125–0.062	Very fine sand
0.062–0.031	Coarse silt
0.031–0.016	Fine silt
0.016–0.008	Very fine silt
0.008–0.004	Coarse clay
0.004–0.002	Coarse clay
0.002–0.001	Medium clay
0.001–0.0005	Fine clay
0.0005–0.00024	Very fine clay

v is the kinematic viscosity of the water

$$w_s = \frac{(s-1)\,gd^2}{18v} \quad 1 \leq d \leq 100\,\mu m \tag{2.2}$$

in which g is the gravitational acceleration, d is the diameter of the particle, and s is the specific density of the particle, namely the ratio of the density of the sediment ρ_s to the density of water, ρ_w as given below:

$$s = \rho_s / \rho_w \tag{2.3}$$

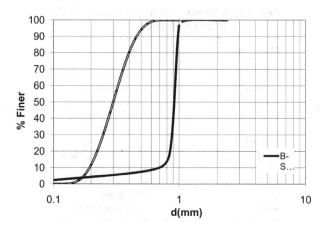

Figure 2.2 Grain diameter distribution of samples for two different sands (A series has $d_{50} = 0.298\,mm$ and B series has $d_{50} = 0.912\,mm$)

For the larger particles the equation is adjusted as follows (Depeweg and Mendez, 2007):

$$w_s = \frac{10v}{1d}\left[\left(1 + \frac{0.01\,(s-1)\,gd^3}{v^2}\right)^{0.5} - 1\right] \quad 100 \leq d \leq 1000\,\mu m \qquad (2.4a)$$

$$w_s = 1.1\sqrt{(1-s)\,gd} \quad d \geq 1000\,\mu m \qquad (2.4b)$$

On the other hand if there are more than one particle falling the group effect should be considered. The fall velocity of a group of particles, is the function of the Reynolds number and the sediment concentration, C (Depeweg and Menendez, 2007).

Any sediment sample taken in the nature is not uniform, therefore a representative diameter should be assigned. The cumulative size–frequency curve which is constructed with the percentage, P_i, of the sediment by mass is smaller than given size, d_i. Mean diameter is computed as:

$$d_m = \sum P_i d_i \bigg/ \sum P_i \qquad (2.5)$$

However, according to the experience gained in the field and in the laboratory, d_{50}, that is 50%, of the sediment finer is the best representative of the sediment sample. The grading of the sample is measured according to the geometric standard deviation:

$$\sigma_g = \sqrt{d_{84}/d_{16}} \qquad (2.6)$$

or the uniformity ratio:

$$C_u = d_{60}/d_{10} \qquad (2.7)$$

If $\sigma_g \leq 1.4$ and $C_u \geq 3.0$ is the sample is accepted as well graded and uniform, respectively.

2.2   MODES OF SEDIMENT TRANSPORT

Sediment particles are transported by flow in one or a combination of ways, bed load, saltation, and suspension. The bed load is jumping, rolling or sliding of the particles on or near the streambed surface creep. Saltation is jumping into the flow and then resting on the bed. Suspension is supported by the stream turbulence surrounding fluid during a significant part of its motion.

If rolling, sliding and jumping characterize the motion of sediment particles it is called bed-load transport (Yang, 1996). On the other hand, suspended-load transport is the motion of sediment particles that is supported by the upward components of turbulent currents and stays in suspension for an appropriate length of time (Yang, 1996). Suspended particles have a diameter of less than 0.062 mm (mainly silt).

Based on previous bed-load and suspended-load transport definition, total-load can be defined as the sum of bed-load and suspended-load. However, based on the source of material being transported, total load can also be defined as the sum of bed-material load and wash load. Wash load consists of fine materials that are finer than those found in the bed. The amount of wash-load depends mainly on the supply from the watershed; not on the hydraulics of river (Yang, 1996). In the light of the definitions, a mathematical representation including bed-load, suspended load, bed material-load and wash-load can be written as:

$$(Total - Load) = (Bed - Load) + (Suspended - Load)$$
$$= (Bed - Material - Load) + (Wash - Load)$$

The ratio of bed-load to suspended load is approximately 5–25%, however, for course materials, a higher percentage of sediment may be transported as bed-load (Yang, 1996).

Sediment transport over movable boundaries of a channel starts if the necessary conditions exceed the critical condition of motion of the bed material (Simons and Şentürk, 1992). The incipient motion has been studied extensively over the past 60 years, following the work by Shields (1936), who presented a semi-empirical approach. Much of the subsequent research into incipient motion builds on the original work of Shields. In addition, only limited comparisons of Shields method against alternative methods for predicting incipient motion have been done. Hjulström (1935) and Yang (1973) studied on expressing the phenomenon in terms of velocity related terms rather than shear parameter, as a function of shear Reynolds number. Those studies mentioned are for the initiation of motion in alluvial streams, i.e. flow with velocity profile of logarithmic distribution law.

Shields (1936) performed several experiments with different ranges of uniformly distributed types of sand. The results of Shields' study were presented as a narrow band with a certain width serving a region below which the tractive force is not sufficient to initiate motion, and above which motion is accepted as already initiated. The tractive force coefficient, θ_*, is plotted as a function of the friction grain Reynolds number, Re_*, as shown in below:

$$\theta_* = f(Re_*) \tag{2.8}$$

$$\frac{\tau_c}{g(\rho_s - \rho)d} = f\left(\frac{u_* d}{\nu}\right) \tag{2.9}$$

in which the shear velocity is defined as:

$$u_* = \sqrt{\frac{\tau_c}{\rho}} = \sqrt{gRS} \tag{2.10}$$

where τ_c is the critical shear stress of the particle. S is the channel longitudinal bed slope and R is the hydraulic radius as defined below:

$$R = A/P \tag{2.11}$$

where A is the flow area and P is the wetted perimeter.

There exist numerous researches in literature which build up additional data points on this diagram. Paphitis (2001) studied the Shields diagram comprehensively including the data added on it from 1914–1994 and defined the borders of the broadened band with equations given below:

Lower limit,

$$\theta_{*cr} = \frac{0.075}{0.5 + Re_*} + 0.0300(1 - 0.699\,e^{-0.015Re_*}) \quad 0.01 < Re_* < 10^5 \tag{2.12a}$$

Upper limit,

$$\theta_{*cr} = \frac{0.280}{1.2 + Re_*} + 0.0750(1 - 0.699\,e^{-0.015Re_*}) \quad 0.01 < Re_* < 10^5 \tag{2.12b}$$

However, Shields diagram and others focused on shallow waters where the logarithmic law of velocity distribution prevails. Thus, the velocity is distributed such that it is zero at the bottom of the channel, increases rapidly towards the free surface and its maximum is often slightly below the free surface (Graf, 1998). However, the flow in the vicinity of an operating bottom outlet of a reservoir, due to larger water depths, complexity of geometry, range of scale and unknown bottom conditions, is very much similar to flow in the vicinity of an orifice as depicted in Figure 2.3. Shammaa et al. (2005) and Kemalli (2009) studied the variation of the velocity profiles with distance from a sluice gate. The velocity distribution is quite different than the logarithmic law of velocity distribution. Even though the velocity distributions for open channels and reservoirs are essentially different, the basic mechanism that leads to initiation of motion remains the same (Gobelez, 2008).

The dimensions of the sediments are relatively small compared to that of the flow parameters; thus, the turbulence will play an essential role in all flows of a mixture consisting of water and sediment. In the case of a reservoir, the dimensions complicate the calculations which lead us to make new definitions other than Shields' parameters.

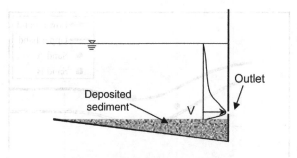

Figure 2.3 Typical velocity profile in front of an outlet

In her study Gobelez (2008) demonstrated a conversion method in order to apply the work of Paphitis (2001) to the problem of the initiation of motion under the effect of a bottom outlet in a reservoir is studied. The procedure is explained as follows:

i) In order to discard the shear velocity, and consequently the use of hydraulic radius, R, and slope, S, out of the parameters, the formula made available by Chezy for the uniform flow is used:

$$V = C_z \sqrt{RS} \tag{2.13}$$

where C: the roughness coefficient of Chezy [$L^{1/2}/T$].

ii) A simple procedure is performed on the Shields' parameters defining the initiation of motion and two well-known parameters of Shields, with the help of constant multipliers, can be re-expressed in terms of the grain Reynolds number and the grain Froude number as:

$$Re = \frac{Vd}{\nu} = \frac{C_z}{\sqrt{g}} Re_* \tag{2.14a}$$

$$Fr^2 = \frac{V^2}{g(s-1)d} = \frac{C_z^2}{g} \theta_* \tag{2.14b}$$

iii) The upper and lower limits, which belong to the initiation of motion band has been defined by Paphitis (2001) are converted with the value of $C_z = 77$. The Chezy coefficient depends on the roughness of the pipe and the hydraulic radius and slope of the open channel (Williams and Hazen, 1911). The variation of C_z is given from 5 for very rough surfaces to 77 for very smooth surfaces in the SI unit system (Heasted Methods, 2003). The bottom of the flow may be assumed as a smooth surface because the dimensions of the flow are very large compared to the ones of the particles to be transported. In this way, the value of the Chezy coefficient is considered from the value representing the smoothest case, starting from 77. Figure 2.4 shows the converted chart of the Paphitis which

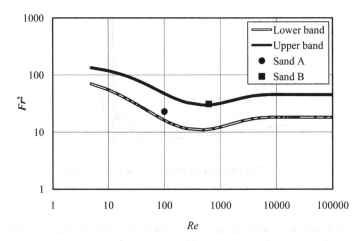

Figure 2.4 Converted Chart of Pahpities with two experiments of Göbelez (2008)

was developed and proofed by two different sand diameters in the study of the Gobelez (2008). In Figure 2.4 the diameter of the Sand A has $d_{50} = 0.298$ mm and Sand B series has $d_{50} = 0.912$ mm.

2.3 SILTATION MECHANISM

Surface erosion and the decomposition of bed-load create suspended-load. Since the settling velocity of the fine material is very low, the particles can sustain in a suspension mode far into the reservoir pool. Some fraction of the fine sediment may be released over the spillway or through the bottom outlet with water, but the most important part is deposited at different locations in the reservoir. Suspended fine sediment accumulates all over the reservoir-space and its amount is generally much larger than that of the coarse sediment. The coarse bed-load deposits mostly in the upper reach of the reservoir pool forming a triangular shaped depositional formation called a delta. Fan and Morris (1992) indicate that delta formation can be divided into two parts as topset bed and foreset bed (Figure 2.5). Relatively finer sediment particles such as silt and clays are carried by the density currents down to the dam structure.

It is not necessary that all the produced sediment on the basin will completely enter into a downstream reservoir. Some faction of it will deposit in natural or man-made barriers and in the channels and their flood plains. The nature and amount of the sediment production in the watershed-area is affected by the amount, intensity and distribution of the rainfall in space and time; soil texture; steepness and slope of the ground surface; vegetation cover and land use (Yılmaz, 2003).

The unit weight of sediments deposited within a reservoir increases in time. The coarse sediment-bank which is situated at the "head" of reservoir, where the river enters into the pool, dries up during the low rainfall season, and new particles are

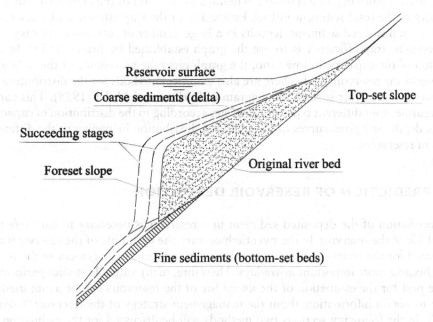

Reservoir surface

Coarse sediments (delta)

Top-set slope

Succeeding stages

Original river bed

Foreset slope

Fine sediments (bottom-set beds)

Figure 2.5 Delta deposition in a reservoir

deposited into the deep fissures of dried soil. The fine sediment becomes more and more compact under the weight of newly deposited sediment (Yılmaz, 2003).

There are several factors influencing the characteristics of sediment deposit in a reservoir as follows: Geometry of the reservoir such as thalweg slope, length and width, sediment properties, trap efficiency of the reservoir, vegetation just upstream of the reservoir head and reservoir operation methods. Within the life of a reservoir three phases can be described from the point of view of the siltation process (Yılmaz, 2003):

i)   The first phase: The capacity of the reservoir is higher than the average annual yield of the catchment area. This phase ends when the reservoir capacity is decreased to the level of the annual discharge of the catchment area. If the reservoir outlets are kept locked the total amount of sediment will be stored in the reservoir pool. Because of this, during this period, the sediment-accumulation grows in a linear-way.

ii) The second phase: The capacity of the reservoir is already smaller than the average annual yield of the catchment area. Among these conditions a certain part of the fine sediment runs over the spillways, even if the reservoir outlets are kept locked. The rate of accumulated sediment will decrease while comparing with the first phase. Some reservoirs are in this stage from the beginning of their operation.

iii) The third phase: The reservoir is completely filled with sediment and the useful life of the dam is finished. Nevertheless the coarse sediment will continue to deposit on the top of the filled-up reservoir while the river tries to re-establish the original slope of the valley.

The trap efficiency of a reservoir is defined as the ratio of the deposited sediment quantity to the total sediment inflow. Estimation of the trap efficiency of a reservoir is based on measured sediment deposits in a large number of reservoirs. An easy way to determine trap efficiency is to use the graph established by Brune (1953). In the function of the capacity–inflow ratio, the graph gives the percentage of the sediment trapped in the reservoir pool. There are also empirical methods for the distribution of deposits in a reservoir such as the diagram of Borland and Miller (1958). This curve distinguishes four different types of reservoirs according to the distribution of capacity versus depth, and gives curves of the sediment distribution belonging to the different types of reservoirs.

2.4 PREDICTION OF RESERVOIR DEPOSITION

The prediction of the deposited sediment in a reservoir is necessary to calculate the useful life of the reservoir. In the twentieth century the useful life of the reservoirs are estimated for the economical analyses. However, sediment management in the reservoirs became more important nowadays. Therefore, many valuable studies performed in the past for the estimation of the useful life of the reservoirs can be again used in order to obtain information about the management strategy of the reservoir (Göğüş, 2007). In the following sections two methods will be discussed for the estimation of the reservoir deposition. The first one is a graphical method, which can be applied to a reservoir in operation for number of years. The second one is mathematical model to predict delta formation. Although it is not capable of representing the simulation of the complete reservoir area, it can be very powerful tool during the planning stage while searching an appropriate reservoir size.

2.4.1 Prediction of the useful life of a reservoir

In any kind of analysis it should be a first step to collect all available data. There are several factors affecting sediment production in a watershed area: amount and distribution of rainfall, soil type and geological formation of the catchment area, topography, geomorphology, ground cover and land use. Also there are several factors influencing the characteristic of the sediment deposit in a reservoir, such as thalweg slope of the reservoir, size and mineralogy of sediment, trap efficiency of the reservoir, vegetation above the reservoir-head and method of reservoir operation. In order to perform any kind of analysis, the first step will be the collection of data. It is quite difficult to get all the data qualitatively and quantitatively.

The two most important hydrological parameters, annual rainfall and surface runoff coefficient, which are necessary to calculate the annual yield of watersheds belonging to the reservoirs can be collected from the map series of the World Water Balance. These maps are not so detailed therefore if there is any value obtained from a deep study, it should be used. Capacity tables of subsequent measurements dealing with the siltation of studied reservoirs should be obtained from the respective agency or organization that is responsible from the dam.

The first step of the prediction process is to collect and compare the results of earlier measurements concerning the storage capacity of the reservoir in question. It is

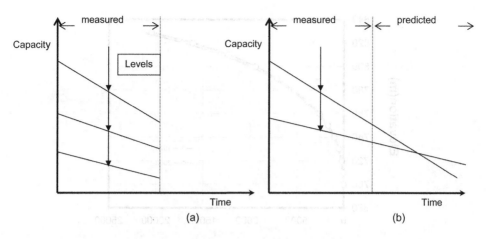

Figure 2.6 Schematic representation of capacity change as a function of time for different levels:
a) Expected and b) Unexpected (Altun et al., 2004)

accepted that the first determination of the capacity is the most precise – when the whole future reservoir-pool was measured among dry conditions by land-survey methods, without hurry. It is also assumed that the first measured capacity remains the same until the end of the construction period (Altun et al., 2004). Any implemented capacity measurements done after the construction are always less accurate when compared to the first one. The most important reason for this is that one part of the survey is accomplished in the water-covered part of the reservoir-pool. To determine the position of the measuring boat, the depth of the very soft surface of the sediment deposition produces a series of problems even in the case of the most modern equipment. In addition the measurement must be finished in a relatively short period.

The second step of the prediction process is to draw elevation-capacity curves. The basic principle that the storage capacity of a reservoir diminishes in every level, and between the levels too, as the sediment is deposited everywhere in the pool. For the purpose of prediction it is more advantageous to represent the change of capacities belonging to different levels as a function of time (Figure 2.6). The curves belonging to the higher levels must always have a more steep slope than those belonging to the lower levels but the slopes must remain proportional (Figure 2.6a). This representation gives us a very strong tool to control the quality of capacity-measurements: e.g. it is impossible that the total (highest-level) capacity will be filled up earlier than a partial (lower-level) capacity (Figure 2.6b).

As the next step of prediction, the average annual yield of the catchment area will be determined. Best results can be obtained by using direct discharge measurements on river stations. Regional hydraulic data can also be used.

If the initial reservoir-capacity is greater than the average annual yield of the watershed, the sediment filling will increase in a linear way. Then, the average silt deposition will be the difference of the initial- and last-measured capacities divided by the years of observation. Then the specific sediment-yield of the watershed ($m^3/km^2/year$) can be calculated by dividing the average annual sediment yield to the catchment area.

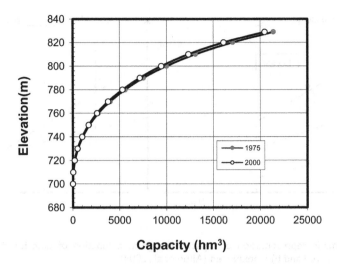

Figure 2.7a Capacity elevation curve of the Keban Dam (Yılmaz, 2003)

If the initial reservoir-capacity is smaller than the discharge of the watershed, the sediment filling will grow in a "decreasing" way (Brune, 1953). In that case the average annual sediment-yield arriving from the catchment area can be calculated as follows.

The amount of the sediment-yield trapped in the reservoir, is given by the difference of the initial- and last-measured capacities. Divided by the years of observation the trapped annual load of sediments can be calculated. The observed trap-efficiency belongs to the capacity/inflow-ratio which was characteristic somewhere in the middle of the observed period, where the tangent of the decreasing curve is parallel to the linear line. Using the approximate capacity/inflow-ratio we can read the trap-efficiency, characteristic for the observed period from the Brune-curves (Brune, 1953). The observed annual amount of sediment (deposited in the reservoir-pool), divided by the trap-efficiency gives the total annual sediment-yield, arriving from the catchment. The specific sediment-yield of the watershed (m^3/km^2/year) can be calculated as described above paragraph.

When the average annual sediment-yield and the type of siltation ("linear" or "decreasing") are determined we have to say something about the size of the arriving sediment-particles. The coarse sediment can be trapped at the head of the reservoir and removed with the advantage of dry methods. It is better if we prevent the coarse sediment from entering into the main reservoir-pool as it can hinder the hydraulic dredging. Coarse sediment arrives only if the initial slope of the reservoir is considerable. The possibility of the coarse sediment deposition can be stated according to the slope of the thalweg (Vituki, 2003). According to Vituki (2003) if the slope of the thalweg is less than 2.9 m/km there is no coarse sediment deposition and if it is larger than 10, there is extreme rate of coarse sediment deposition.

Figure 2.7 shows the capacity-elevation curve of two dams one is a large dam (Figure 2.7a) and the other is a reletively small dam (Figure 2.7b). Table 2.2–2.5 shows the results of the predictions of the six dams with the graphical method explained above. Table 2.2 shows the general properties of the selected six reservoirs. Table 2.3

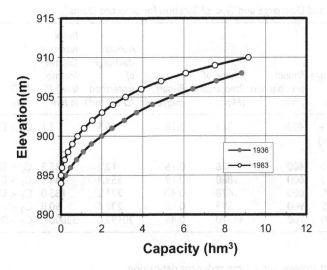

Figure 2.7b Capacity elevation curve of the Çubuk Dam (Yılmaz, 2003)

Table 2.2 Physical Properties of Selected Dams

Name	Starting year	Completion year	River	Province	Embankment type	Dam volume (m³)	Purpose
Cubuk I	1930	1936	Cubuk	Ankara	Concrete Gravity	120000	Water Supply + Flood Cont.
Cubuk II	1961	1964	Cubuk	Ankara	Earthfill	1100000	Water Supply
Kemer	1965	1958	Akcay	Aydin	Concrete Gravity	740000	Water Supply + Flood Cont. + Energy
Kartalkaya	1965	1972	Aksu	K.Maras	Earthfill	1452000	Irrigation + Water Supply
Hasanlar	1965	1972	K.Melen	Bolu	Rockfill	1651000	Irrigation + Flood Cont.
Keban	1965	1975	Euphrates	Elazig	Concrete Grav. + Rockfill		Energy

shows watershed discharge and type of siltation. The last column gives the information related to the trend of the decreasing capacity. Table 2.4 shows the slope of thalweg and the possibility of coarse sediment deposition. It can be seen that in the case of one reservoir (Keban) there is no need to fear the problem of the coarse-sediment deposition, but for the others precautions must be made. Table 2.5 is the predicted half life time of the reservoirs and estimated sediment yield of watershed. The date when the initial storage-capacity is degraded to its half value is a very important date in the span of a reservoir's life. It is the time to plan the substituting installations, to change the management-concept or to reduce the needs.

Table 2.3 Watershed Discharge and Type of Siltation for Selected Dams

Name of the reservoir	Drainage Area (km^2)	Annual precipitation (mm)	Volume of Precipitation (Hm^3)	Runoff coefficient	Annual discharge of watershed (D in Hm^3)	Initial reservoir Capacity (normal level) (C_0 in Hm^3)	C_0/D ratio	Type of siltation at the beginning
Cubuk I	(720 + 190 =) 910	400	364	0.18	65.5	7.1	$C_0 < D$	Decreasing
Cubuk II	190	400	76	0.16	12.2	25.3	$C_0 > D$	Linear
Kemer	3100	600	1860	0.19	353.4	544	$C_0 > D$	Linear
Kartalkaya	1130	600	678	0.40	271.2	185.0	$C_0 < D$	Decreasing
Hasanlar	665	900	599	0.40	239.6	50.8	$C_0 << D$	Decreasing
Keban	64100	700	44870	0.45	20192.0	32600	$C_0 > D$	Linear

Table 2.4 Slope of thalweg and coarse sediment deposition

Name of the reservoir	Dam crest elevation (m)	Height of dam above thalweg (m)	Elevation of river bed at dam-site (m)	Normal Level (m)	Initial depth of reservoir (m)	Length of reservoir (km)	Slope of thalweg (m/km)	Expected coarse-sediment deposition
Cubuk I	908.6	25.0	883.6	906.6	23.0	5	4.6	◇
Cubuk II	1117.0	61.0	1056.0	1113.0	57.0	3	19.0	◇◇◇
Kemer	298.5	108.5	190.0	291.5	101.5	17	6.0	◇◇
Kartalkaya	722.0	56.0	666.0	715.7	49.7	13	3.8	◇
Hasanlar	272.8	70.8	202.0	255.5	53.5	8	6.7	◇◇
Keban	848.0	163.0	685.0	845.0	160.0	125	1.3	−

Table 2.5 Predicted values and estimated sediment yield of watershed

Name of the reservoir	Half life time (years)	Expected date of $C_0/2$	Expected date of total filling up	Specific sediment yield of watershed ($m^3/km^2/year$)
Cubuk I	68	2004	2090–2120	83
Cubuk II	150	2114	Around 2270	447
Kemer	62	2020	2083–2088	1600
Kartalkaya	45	2017	2063–2066	1840
Hasanlar	56	2028	2095–2110	722
Keban	235	2210	After 2450	997

The length of the "half life-time" depends also from the ratio of C_0/D. If a large reservoir is built, which is mostly empty, because the watershed discharge is not enough to fill it ($C_0/D >> 1$) the "half life-time" will be long. This helps the goals of the concept concerning the guaranteed hydraulic situation, but the dam-openings are in constant danger of siltation. The proper C_0/D ratio must be determined by optimization.

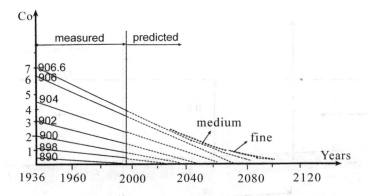

Figure 2.8 Capacity Prediction of Cubuk Dam (Altun et al., 2004)

Values of "half life-time", the year when the capacity presumably reaches the $C_0/2$ value, predicted dates of total siltation and specific sediment-yield of the watersheds belonging to the studies are given in Table 2.5.

Figure 2.8 is one of the prediction diagrams developed. In this diagram measured capacities over 20 years are drawn for each level. Then specific sediment-production rate of the watershed is calculated.

2.5 MATHEMATICAL MODELS

2.5.1 Saint-Venant-Exner equations

Simplified longitudinal illustration of water and sediment discharges entering and leaving a reach is illustrated in Figure 2.9 in which q and q_s is the water discharge and the sediment discharge, respectively. S is the slope of the channel, S_w is the water surface slope, S_e is slope of the energy grade line. x is the longitudinal distance representing the flow direction, z is the coordinate along the gravitational directions and h is the flow depth. The equation of continuity for a one-dimensional open channel flow in a prismatic channel is expressed as:

$$h\frac{\partial U}{\partial x} + U\frac{\partial h}{\partial x} + \frac{\partial h}{\partial t} = 0 \tag{2.15}$$

where U is flow velocity and t is time. The Dynamic Equation for an unsteady and gradually varied flow is given by:

$$\frac{1}{g}\frac{\partial U}{\partial t} + \frac{U}{g}\frac{\partial U}{\partial x} + \frac{\partial h}{\partial x} + \frac{\partial z}{\partial x} = -S_e \tag{2.16}$$

where g is gravitational acceleration and S_e is the slope of the energy grade line that is a function of friction factor, f, U and h.

$$S_e = \text{func.}(f, U, h) \tag{2.17a}$$

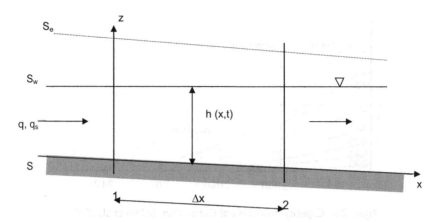

Figure 2.9 Simplified representation of a channel reach

In general S_e calculated from Manning Equation.

$$q = \frac{1}{n} y R^{5/3} \sqrt{S_e} \qquad (2.17b)$$

The Mannings' n is given with the Manning Strickler Equation (Garcia, 2007):

$$n = \frac{k_s^{1/6}}{8.1 g^{1/2}} \qquad (2.17c)$$

k_s is the roughness height.

For flow over mobile beds, i.e., the channel bed elevation change depends on x and t, Exner (1925) developed an erosion equation expressed and Exner's erosion equation is written by Graf (1971) as a continuity equation for solid phase as:

$$\frac{\partial z}{\partial t} + \frac{1}{1-p} \frac{\partial q_s}{\partial x} = 0 \qquad (2.18)$$

where p is porosity defined as the ratio of the volume of pore space which is filled with water and air to the total volume. q_s is the sediment discharge per unit width, i.e. flow of solid particles depending upon U and h of the water sediment mixture (Graf and Altınakar, 1998):

$$q_s = \phi(U, h) \qquad (2.19)$$

Three of the Equations of Saint-Venant-Exner given above (2.15–2.17) are the expressions of the liquid phase of the water sediment mixture flow. The other two equations (2.18–2.19) are equations of solid phase. Since, equations for both phases include implicitly the semi-empirical equations, simultaneous solution becomes too complicated if these equations are applied for movable bed sediment transport (Graf and Altınakar, 1998).

In order to determine the bed level variations various analytical and numerical methods were developed for the solution of Saint-Venant-Exner equations. For the case of analytical solutions severe assumptions have to be made such as steadiness and uniformity of the flow at the same time. Nonlinear solutions of aggradation and degradation for movable bed were given by Gill (1987). If the flow is unsteady and nonuniform or steady and nonuniform there is no available simple solution. In that case time consuming solutions of Saint-Venant-Exner must be done numerically by the help of computers (Graf and Altınakar, 1998).

2.5.2 Developed methods for the solution of Saint-Venant equations

There are various models in literature aiming to solve for bed level variations especially proposed for the river-reservoir system.

Chen et al. (1978) model employs numerical solution of Saint-Venant-Exner equations. In their model, reservoir has been taken as laterally compound so that the effect of changes in lateral direction, i.e. variations in width of channel-reservoir system on bed level changes can be investigated.

Holly and Rahuel (1990) proposed a new numerical physical framework for mobile bed modelling in order to overcome the inadequate assumptions in existing models such as quasi-steady flow assumption. This assumption does not allow an annual hydrograph to be presented in the model. Also, in the proposed model a distinct separation of suspended and bed-load is used via fully coupled numerical computation scheme in other words, synchronous solution of the equations of unsteady water and sediment movement.

A two dimensional vertical model for reservoir sedimentation was developed by Tarela and Menendez (1999) which is also capable of accounting some three dimensional effects due lateral integration of equations of motion. This numerical model also employs the asynchronous i.e. decoupled numerical solution of hydrodynamic and sediment concentration equations.

Toniolo and Parker (2003) proposed a numerical model which considers the depositional behaviour of sand and mud in a reservoir in two parts: delta deposition was calculated by employing numerical solution of Saint-Venant-Exner equations and the density current calculation.

2.5.3 Developed computer models for aggradation and degradation processes

Hydrologic Engineering Centre (HEC) of U.S. Army Corps of Engineers developed a one dimensional computer model called HEC-6 to study the scour and deposition in rivers and reservoirs (USACE, 1977 and 1993). HEC-6 model does not have the capability of simulating bed forms; however, it includes the resistance of bed forms by a variable Manning's Roughness Coefficient, which may change with discharge (Tingsanchali and Supharatid, 1996). Channel geometry, hydraulic data, sediment and hydrographic data are the inputs of the model to calculate flow velocity, water level, bed levels and sediment transport rate.

The HEC-6 model, considers the interaction of water flow and sediment transport. In the hydraulic part two governing equations are employed namely the continuity equation for quasi-steady flow:

$$Q = V \cdot A \tag{2.20a}$$

Momentum equation for quasi-steady non-uniform flow is:

$$S_e = S - \frac{\partial h}{\partial x} - \frac{\partial}{\partial x}\left(\frac{U^2}{2g}\right) = \frac{n^2 V^2}{R^{4/3}} \tag{2.20b}$$

where, S_e friction slope, S bed slope, n Manning Roughness coefficient, R hydraulic radius and V are average flow velocity.

On the other hand, for the sediment part two governing equations are used namely, the continuity equation for sediment transport and sediment transport equations as given below, Eq. 2.21 (USACE, 1977 and 1993):

$$\frac{\partial q_s}{\partial x} + B\frac{\partial z}{\partial t} = 0 \tag{2.21a}$$

$$q_s = \phi(U, h, d, n, ...\text{etc.}) \tag{2.21b}$$

The backwater profile calculation is performed by standard step method. For sediment part, equation is transformed into a finite difference equation using a central explicit finite difference scheme. After bed-level changes calculations, model continues the computation for the new backwater profile for a given time interval and location (Tingsanchali and Supharatid, 1996).

Molinas and Yang (1986) developed a model for the U.S. Bureau Reclamation which is called Generalized Stream Tube Model for Alluvial River Simulation (GSTARS). The stream tube concept is used in GSTARS to provide a semi-two dimensional simulation of flow conditions with a one dimensional approach along stream tubes (Yang, 1996). For the computation along each stream tube and optimization method is applied which is called Minimum Unit Stream Power Theory developed by Yang (1973). According to the theory, an alluvial river tends to adjust its channel geometry and slope in order to minimize its unit stream power since it is the rate of potential energy dissipation per unit weight of water.

The numerical method employed in GSTAR models (for versions GSTARS, GSTARS 2.0, GSTARS 2.1 and GSTARS 3.0) is a finite difference asynchronous method that at first computes water surface profile then performs sediment routing and computes bed level changes by keeping all the hydraulic parameters constant during calculations. GSTARS has four basic sediment transport equations for user to choose. These sediment transport equations are Yang's (1973) sand equation and gravel equation proposed in 1984, Ackers and White's (1973) equation and Engelund and Hansen's (1972) equation. In addition to these sediment transport equations the program user has the chance to add the program code of his/her choice in order to compare the performance of the equations.

Yücel and Graf (1973) developed a model which is for prismatic channels of unit width and for uniform bed slope. Three different bed-load equations were used in the model namely, Einstein (1942), Schoklitsch's (1950) and Meyer-Peter and Müller (1948) equation. Backwater calculations are done by a step method for a constant discharge and uniform grain size of three different diameters (0.5 mm, 1 mm, 2 mm) were used. Although the method was strictly applicable to prismatic channels only, computer applications resulted with patterns of delta formation and progression that are similar to those observed in existing reservoirs, such as the delta of the Lake Mead, behind Hoover Dam (Durgunoğlu, 1980).

Durgunoğlu (1980) developed another computer model which is also based on the uncoupled solution of equations of backwater profile and a sediment transport equation. Backwater profile computations are done by standard step method. For the sediment transport capacity and deposition calculations only one equation, Einstein 1950 equation, is coded as a subroutine in the main program. Employment of the Einstein (1950) equation brought the user the opportunity of executing the computer model for effects of different particle sizes determined by a granulometric distribution curve rather than a uniform average particle size. Another feature of this model is its capability of calculating for non-prismatic channels. Computer applications of the Durgunoğlu (1980) model showed that the location of the delta deposit is affected strongly by the changes in the granulometric curve. However the location of the delta deposit is not affected very much by the variations in channel cross-sections.

In order to calculate the sediment deposition in reservoir–river systems, in other words, to simulate the cycle of modification of bed and resulting backwater profile Graf and Altınakar (1998) developed a computer program called DELTA. DELTA was written in standard FORTRAN programming language which is especially capable of calculating deposition of the relatively coarser sediments in the form of a delta.

Pulcuoğlu's (2009) study of qualitative and comparative investigation on the performance of 32 sediment transport equations was made with the objective of determining the effects of different sediment size ranges and river bed slope on the extent, location and height of the sediment deposition in reservoirs. Sediment transport equations, which were grouped according to the median particle size ranges on which they were developed, were added to DELTA, a simulation program, and corresponding computer runs were done.

Table 2.6 is classified according to type of the sediment transport mode 1) bed-load 2) total-load. The table covers a time period of 124 years from 1879 to 2003. However, due to differences in the range of particle size covered it is not possible to apply all 32 equations to a specific problem. Some of the Sediment Transport Equations are directly dependent on the flow depth and average flow velocity while some others depend indirectly via slope of energy grade line.

In the next section the important outcomes of Pulcuoğlu (2009) will be given. Further the flowchart and the code of Pulcuoğlu which is based on the program DELTA of Graf and Altınakar, (1998) are given in APPENDIX. A brief introduction of Program DELTA will be introduced in order to understand the study of Pulcuoğlu (2009). It is suggested to readers to see original work of Graf and Altınakar (1998) in order to understand theoretical concept.

Table 2.6 Sediment transport equations according to sediment particle size ranges (after Pulcuoğlu, 2009)

		Equation Type	Particle Size Range	d_{50} (mm)	Reference
1	Schoklitsch's Approach (1950)	Bed-load	Fine Gravel–Medium Gravel	≥6.0	Graf and Altınakar, 1998 (Bathurst et al, 1987)
2	Meyer-Peter et al. Equation (1948)	Bed-load	Very Fine Gravel–Medium Gravel	>2.0	Graf and Altınakar, 1998
3	Einstein's Equation (1942)	Bed-load	Coarse Sand–Coarse Gravel	0.785–28.65	Gomez and Church (1989) (Einstein 1942)
4	Inglis-Lacey Formula (Inglis-1968)	Total-load	Medium Sand	–	Vanoni (1975) (Vanoni, Brooks and Kennedy, 1960)
5	Engelund and Hansen's Approach (1967)	Total-load	Fine Sand–Coarse Sand	>0.15	Yang (1996),Vanoni (1975)
6	DuBoys' Formula (1879)	Bed-load	Fine Sand–Very Fine Gravel	0.125–4.0	Yang (1996), Gomez and Church (1989)
7	Kalinske's Equation (1947)	Bed-load	Coarse Sand–Coarse Gravel	0.785–28.65	Graf (1971) (Kalinske, 1947)
8	Shields' Equation (1936)	Bed-load	Very Fine Gravel	1.70–2.5	Shields (1936)
9	Laursen's Approach (1958)	Bed-load	Very Fine Silt–Very Fine Gravel	0.011–4.08	Vanoni (1975)
10	Ackers and White's Formula (1973)	Total-load	Medium Sand–Very Fine Gravel	0.04–2.5	Ackers and White (1973)
11	Shen and Hung's Method (1972)	Total-load	Very Fine Sand–Coarse Sand	0.13–1.3	Shen and Hung (1972)
12	Yang's Sand Formula (1973)	Total-load	Sand	<2.0	Yang (1996)
13	Rottner's Formula (1959)	Bed-load	Sand	–	Greystone (1998) (Reid and Dunne, 1996)
14	Toffaleti Procedure (1969)	Total-load	Fine Sand–Coarse Sand	0.3–0.93	Vanoni (1975)

15	Chang, Simons and Richardson's Approach (1967)	Bed-load	Fine Sand–Very Coarse Sand	0.19/0.33/0.36/ 0.50/0.52/0.93	Simons and Şentürk (1992)
16	Bagnold's Equation (1966)	Total-load	Medium Sand–Very Coarse Sand	0.25–2.00	Yang (1996) (Bagnold, 1966)
17	Wu, Wang and Jia (2000)	Bed-load	Very Fine Sand–Very Coarse Gravel	0.073–64.0	Wu, Wang, Jia (2000)
18	Brownlie's Method (1981)	Total-load	Very Fine Sand–Very Fine Gravel	0.088–2.8	Brownlie (1982)
19	Parker's Approach (1982)	Bed-load	Coarse Gravel	18.0–28.0	Yang (1996), Gomez and Church (1989), Parker (1990)
20	Einstein and Brown Approach (1950)	Bed-load	Coarse Sand–Coarse Gravel	0.785–28.65	Simons and Şentürk (1992)
21	Yalin's Formula (1963)	Bed-load	Medium Sand–Coarse Gravel	0.315–28.65	Yalin (1977)
22	Engelund and Fredsoe's Equation (1976)	Bed-load	Coarse Sand–Medium Gravel	0.93–7.76	Engelund and Fredsoe (1976), Nakato (1990)
23	Van Rijn's Formula (1984)	Bed-load	Fine Sand–Very Coarse Sand	0.2–2.0	Van Rijn (1984-Part I [Bedload])
24	Dou's Equation (1977)	Bed-load	Medium Sand–Fine Gravel	0.35–7.02	Dou (1989) (Dou 1977)
25	Karim and Kennedy Approach (1990)	Total-load	Coarse Silt–Coarse Gravel	0.08–28.60	Karim and Kennedy (1990)
26	Bishop, Simons and Richardson Method (1965)	Total-load	Fine Sand–Coarse Sand	0.19/0.27/ 0.47/0.93	Simons and Şentürk (1992)
27	Wilcock and Crowe Method (2003)	Bed-load	Coarse Sand–Very Coarse Gravel	0.5–45.3	Wilcock and Crowe (2003)
28	Einstein Method (1950-Bed-load)	Bed-load	Coarse Sand–Coarse Gravel	0.785–28.65	Einstein (1950)
29	Colby Relation (1964)	Total-load	Fine Sand–Coarse Sand	0.1–0.8	Yang (1996)
30	Yang's Gravel Formula (1984)	Total-load	Very Fine Gravel–Medium Gravel	0.2–10.0	Yang (1996)
31	Yang and Lim's Formula (2003)	Total-load	Medium Sand–Very Fine Gravel	0.833–3.54	Yang and Lim (2003)
32	Molinas and Wu's Equation (2001)	Total-load	Coarse Silt–Coarse Sand	0.091–1.15	Wu and Molinas (2001)

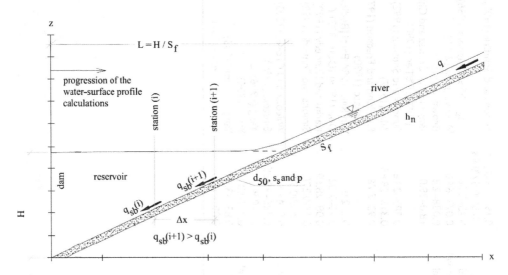

Figure 2.10   River-Reservoir system for which DELTA is developed (modified from, Graf and Altınakar, 1998)

2.5.3.1   Prediction of delta formation

When a dam is constructed on a river it will have a longitudinal profile like the one shown in Figure 2.10. The dam creates a backwater curve that must be calculated in order to determine the hydraulic parameters, such as average velocity, water depth, slope of energy grade line, etc., for the whole length of the system. For the stations, (i) and (i + 1) shown in Figure 2.10 if both of the hydraulic parameters and sediment properties are known, sediment carrying capacities can be calculated. Sediment carrying capacities of flow through to successive stations are calculated with sediment transport equations that are given in Table 2.6. As a consequence, the change in the sediment transport capacity due to the hydraulic conditions, determined from backwater profile calculations, causes erosion or deposition in a reach between two adjacent stations of Figure 2.10.

Although the cross-sectional geometry of natural rivers have complex forms, the hydrodynamic equations and bed-load transport equations are adopted for a unit width since the length of a river-reservoir system is much larger than the depth of flow (Graf and Altınakar, 1998). For water surface computations a dam constitutes a control section i.e., boundary section. The flow regime is subcritical and the flow depth is known, thus the calculations start at the dam and proceed towards the upstream (Graf and Altınakar, 1998). The length of the reach of reservoir-river system can be determined by estimating the upstream end where the bed-load transport is in equilibrium and river attains its normal depth. Thus in order to define the upstream boundary a sufficiently long river reach should be chosen. Normal depth of flow is calculated from the Manning equation for unit discharge and assuming wide rectangular channel (Graf and Altınakar, 1998).

The approximate length of the reservoir can be approximated by dividing the water depth at the dam, by energy grade line slope. However, attention must be given to the length of computational domain. In order to guarantee a uniform flow during whole simulation, the length of the computational domain should be chosen much greater than the reservoir length. The computational step is obtained by dividing the length of the computational domain in the number of reaches (Graf and Altınakar, 1998).

With known bed elevations calculations start at time $t = 0$. At first, calculations for hydraulic parameters are carried out to determine the water surface profile. Once the hydraulic parameters are determined at each station, bed-load transport rates are calculated for all stations by using one of the sediment transport equations mentioned previously. Sediment transport capacity differences give the deposition or erosion rate for reaches between adjacent stations. After calculation of sediment volumes of reaches for a time interval Δt, sediment deposition heights are calculated. Calculation of deposition height for each section is the end of computation for the initial time ($t = 0$). By considering the modified bed elevations computation continues for next time interval, $t = \Delta t$, up to a specified duration of simulation. The program DELTA is based on a decoupled, asynchronous algorithm which means that the calculations for the liquid and solid phase are carried out separately and successively; during the calculations for one phase, the characteristics of the other phase are kept constant (Graf and Altınakar, 1998).

In his study Pulcuoğlu (2009), kept the original form of the program code of DELTA except for adding codes of 29 subroutines of sediment transport equations. In the origin of the DELTA programme in Table 2.6 the first three equations are used. In addition to that, explanations of the variables for each of the sediment transport equations are included with each subroutine added to the main program code. The modular structure of the program is composed of the main program DELTA and nine subroutines, each coded for a specific task: DREAD; TITLES; RK4; DERIVE; SCHOKL; MEYPET; EINS42; FORMUL and DWRITE.

The main program, DELTA, first calls the subroutine DREAD in order to invite the user to input the program data via an interactive dialog screen. Then DREAD carries the program data to the main program, DELTA. Input data that must be given by the user is composed of 6 parts namely:

Physical characteristics: Initial bed slope, average sediment diameter, Manning-Strickler coefficient, densities of the water and of the sediments, discharge per unit width.

As the user starts the program, the initial bed slope and the average diameter of the sediments will be asked. By using an average grain diameter the program calculates the Manning-Strickler coefficient, due to grain roughness, and then prints the value on the screen. Later on, the user is asked the total Manning-Strickler coefficient. If it is thought that the head loss due to irregularities is greater than the loss due to grain roughness, the estimated total value should be entered. Afterwards the input part goes on with the entry of density of sediments and density of water and discharge per unit width (Graf and Altınakar, 1998).

Choice of the bed-load transport formula: Selection of the number assigned to the desired sediment transport formula. The assigned numbers for the subprograms are 1) SCHOKL; 2) MEYPET; 3) EINS42; 4) INGLAC; 5) ENGEHAN; 6) DUBOY;

7) KALIN; 8) SHIELD; 9) LAURSEN; 10) ACKWHI; 11) SHUNG; 12) YANG; 13) ROTTNER; 14) TOFFALETI; 15) CHANG; 16) BAGNOLD; 17) WUWANG; 18) BROWNLIE; 19) PARKER; 20) EINSBROWN; 21) YALIN; 22) ENGEFRED; 23) VANRIJN; 24) DOU; 25) KARKEN; 26) BISHOP; 27) WILCR; 28) EINS50; 29) COLBY; 30) YANGG; 31) YANGLIM; 32) WUMOLIN.

By the addition of subroutines of 29 sediment transport equations, the program now asks the user to choose a number assigned to a bed-load equation among them. For example, in order to perform a simulation using Laursen's equation the user has to choose number 6.

Data concerning the modification of the bed profile: Maximum relative bed-level change tolerated porosity of the sediments, the ratio of the upstream/downstream heights of the sediment deposition or erosion.

First one is asked the maximum relative variation of the bed level for a specified time step. According to this input, program checks during simulation whether the bed level modification is exceeded or not, by comparing the ratio of bed level modification to water depth with the specified the maximum relative variation of the bed level value (Graf and Altınakar, 1998). In order to transform the volume of the sediments deposited into bed level modification heights the ratio between the upstream and the downstream heights of trapezoidal deposits value should be entered. The recommended value is between 0.5 and 1. So that it should be noted that the deposition at the downstream end of a reach is higher than the one at the upstream end (Graf and Altınakar, 1998).

Information concerning the computational domain: Coordinates of the first and the last station, space length, maximum tolerated dynamic head variation, maximum number of subdivisions.

After that the user should enter the information related to the calculation domain. That is to enter the coordinates of the first station and the total distance that ends at the upstream. The origin of the coordinate system is located at the dam. The step length, which is the length between consecutive stations, is important for the total calculation time. If step length is too small the time step must also be small. However this increases the total calculation time. If a large step length is chosen, it may cause errors in the calculation of the water surface profile especially around the region where the river meets the reservoir (Graf and Altınakar, 1998).

Boundary conditions: water depth at the downstream end.

As boundary conditions, the user should enter a value for water depth at dam section. Water surface calculations start from this value.

Simulation time and the printing of results: Time step, duration of the simulation, frequency of the printing of the results and the name of the output file (Graf and Altınakar, 1998).

The last part of the inputs is the specification of the time step, and the total simulation period. For instance if the time step and total simulation period is chosen 10 days and 36500 days (100 years), respectively. The program calculates the 10 day deposition for each reach of the river-reservoir system up to 36500 days. Furthermore, the time interval for the reprinting frequency of results in the output file in terms of number of steps can be specified. If user specifies this number as 730, the calculation results will be written at 20 year intervals.

After DREAD's reading of the program input by questioning the user, subroutine TITLES is called by the main program. TITLES' function is printing the program data on the output file in a tabular form supplied by the subprogram DREAD (Graf and Altınakar, 1998).

The function of the subprogram RK4 is the calculation of water surface profile by using 4th order Runge-Kutta Method. To maintain sufficient precision of the water surface profile calculations, the program asks the user to specify a value for tolerable dynamic head difference, $(\Delta V^2/2g)$, between two adjacent stations. The program then checks the dynamic heads calculated, whether they are less than the tolerated value or not. If any difference of dynamic heads is greater than the limit value, the program subdivides the reach into 2 parts that are defined by corresponding successive stations. If the criteria are still not met the program makes a second subdivision, as $2^2 = 4$ parts, and so on. The criterion for subdivision is also specified by the user up to $2^7 = 128$ subdivisions for a reach. If the criteria are still not met by using 128 subdivisions, then the program stops the run and asks for a smaller step length to be entered (Graf and Altınakar, 1998).

The differential equation for the free surface flow in a rectangular channel is given for a constant unit discharge, $q = Q/B$:

$$\frac{dh}{dx} = -\frac{S - \dfrac{q^2 \cdot n^2}{h^{10/3}}}{1 - \dfrac{q^2}{g \cdot h^3}} \tag{2.22}$$

which is to be solved by subroutines RK4 and DERIVE. At this step calculation for the liquid phase is finished for one time step (Graf and Altınakar, 1998).

After completing the liquid phase calculations for one time step, computation continues with solid phase calculations by corresponding subroutine of the sediment transport equation that is selected at the data input part. Because of the decoupled character of the algorithm, it is assumed that water surface profile does not vary during the calculation of solid phase (Graf and Altınakar, 1998).

The main program, DELTA, takes the calculated values of sediment transport capacities for each station. The quantity of the sediments to be deposited or eroded in a reach, Δq_s, depends on the difference between the bed-load transport capacities at the upstream, $q_s(i+1)$, and at the downstream, $q_s(i)$, stations: $\Delta q_s = q_s(i+1) - q_s(i)$. It should be noted here that the number of a reach is the same as the number of the station at its downstream end (Graf and Altınakar, 1998).

Thereafter, calculating the sediment transport capacities between consecutive stations and deposition (or erosion) rate for corresponding reach, DELTA converts this rate into a deposition height. The volume of the sediments to be deposited at the reach (i), taking into account the volume increase due to the porosity, p, is calculated as:

$$\text{Deposition Volume for the reach (i)} = \Delta q_s(i) \cdot \Delta t \cdot \frac{1}{1-p} \qquad (2.23)$$

During data input the user is also asked to specify a value for the maximum relative variation of bed level. At the end of each time step the program searches for the maximum relative variation of bed level and compares it with the specified value of it (Graf and Altınakar, 1998). If the specified value is less than the relative variation, then the program stops the run and gives a message requesting a smaller time step or a greater maximum relative variation value to be entered.

As a time step ends, if the printing time specified by the user has passed, the subroutine DWRITE is called by the main program for printing the results for the corresponding time step on the output file. If the simulation duration specified by the user is not finished, the program continues the run by modifying bed level heights for each station. Then the program goes to the beginning of the calculation loop to start a new time step with calculating the water surface profile by using new bed profile (Graf and Altınakar, 1998).

2.5.4 Application of DELTA

In order to compare the different sediment transport equations one input set is prepared (Table 2.7). However, the particular diameter range of each equation is not the same as those depicted in Table 2.6.

Therefore, five different median sizes are chosen: 6.0 mm, 2.0 mm, 1.0 mm, 0.5 mm and 0.2 mm. Table 2.8 summarizes the calculations for bed-load equations. In this table the starting point of the foreset bed, uppermost point of the topset bed and the height of the delta are given. In addition for 20 years of simulation the average value and the standard deviation of the group are given. The columns under the discrepancy ratios show the deviation from the mean value. The same information is given for the total load equations in Table 2.9.

The effect of the median size change, the effect of the river slope change and discharge on deposition phenomenon could be investigated by using the programme.

Change in the deposition rate due to the flow discharge can be found out from the deposition percentage graphs. Another set is prepared which is given in Table 2.10 which has different bed slope values.

From different combinations of two input sets, several conclusions are made by Pulcuoğlu (2009). Table 2.11 and Table 2.12 show the deposition percentage change due to decrease in particle size of bed-load run groups and of total-load run groups respectively.

Table 2.13 shows the deposition percentages with respect to the change of unit discharge of bed-load equations and deposition percentage change due to the river slope and discharge changes given in Table 2.14.

Table 2.7 Input table of CASE I

Symbol	Explanation	Variable name in DELTA	Input value	Unit
Physical Characteristics Data				
S	Initial bed slope of the River	SF	0.00054	–
d_{50}	Average Sediment Particle Diameter	D50	*	mm
n_M	Manning-Strickler Coefficient	CN	0.032	$m^{-1/3}s$
ρ_s	Density of Sediment Particles	ROS	2650	kg/m^3
ρ	Density of Water	ROE	1000	kg/m^3
q	Water Discharge per Unit Width	QU	2.5	$m^3/s/m$
Data Concerning the Modification of the Bed Profile				
$\Delta z(i)/h_i$	Maximum Relative Bed-level Change Tolerated	VARZMX	0.1	–
p	Porosity of the Sediments	POROS	0.3	–
λ	The Ratio of the Upstream/Downstream Heights of the Sediment Deposition or Erosion.	HAMHAV	0.75	–
Information Concerning the Computational Domain				
x_I	x coordinate of the first station	XI	0	m
x_f	x coordinate of the last station	XF	120000	m
Δx	Step Length in x direction	DX	600	m
$\Delta V^2/2g$	Maximum Tolerated Variation of Dynamic Head	DHDYNM	0.01	m
–	Maximum Number of Subdivisions in powers of 2	NMC	7	–
Boundary Conditions				
h_I	Water Depth at the Downstream End	HI	23.5	m
Simulation Time and Printing of Results				
Δt	Time Step	DELT	**	days
–	Duration of Simulation	TFIN	7300	days
–	Step Number end of which the results are printed	NPP	***	–

*Calculations were done for 5 sediment particle diameter: 6.00, 2.00, 1.00, 0.5 and 0.2 mm.
**Input Value depends on the selected sediment transport equation and the deposition rate of sediment particles
***Input value depends on the choice of the time step

After simulation with the DELTA program, grouped sediment transport equations were compared with each other according to the average values related to the formation of delta deposits. Besides that, performances of the equations on the deposited volumes were compared with respect to the initial volumes of reservoirs. According to the comparison of simulation results, with respect to the deposition percentages, the following sediment transport equations gave the closer values to the mean values; Einstein Equation (1942), Rottner's Formula (1959), the Karim and Kennedy

Table 2.8 Calculation Results of Bed Load Equations

		Starting Point of Foreset Bed	Uppermost Point of Topset Bed		Discrepancy Ratios		
Equation	Year	Dist. from the Dam (m)	Dist. from the Dam (m)	Height (m)	Starting Point	Uppermost Point	Height of Delta
1 Schoklitsch's Approach (1950)	0 10 20	– 38400.00 37200.00	– 39600.00 38400.00	– 6.003E−01 9.077E−01	1.28	1.26	0.18
2 Meyer-Peter et al. Equation (1948)	0 10 20	– 37800.00 36600.00	– 39000.00 37800.00	– 8.195E−01 1.233E+00	1.26	1.24	0.25
3 Einstein's Equation (1942)	0 10 20	– 30600.00 27600.00	– 33600.00 29400.00	– 3.515E+00 5.397E+00	0.95	0.97	1.09
4 Kalinske's Equation (1947)	0 10 20	– 27600.00 23400.00	– 28800.00 24600.00	– 5.632E+00 7.907E+00	0.80	0.81	1.60
5 Wu, Wang and Jia (2000)	0 10 20	– 21600.00 15600.00	– 24600.00 18000.00	– 8.116E+00 1.168E+01	0.54	0.59	2.36
6 Einstein and Brown Approach (1950)	0 10 20	– 32400.00 29400.00	– 33600.00 30600.00	– 3.304E+00 4.873E+00	1.01	1.01	0.98
7 Yalin's Formula (1963)	0 10 20	– 37800.00 37200.00	– 39000.00 38400.00	– 5.254E−01 9.200E−01	1.28	1.26	0.19
8 Engelund and Fredsoe's Equation (1976)	0 10 20	– 29400.00 25800.00	– 30600.00 27000.00	– 4.747E+00 6.727E+00	0.89	0.89	1.36
9 Dou (1977)	0 10 20	– 36600.00 35400.00	– 37200.00 36000.00	– 1.183E+00 1.871E+00	1.22	1.18	0.38
10 Wilcock and Crowe Method (2003)	0 10 20	– 31200.00 28200.00	– 32400.00 29400.00	– 3.795E+00 5.391E+00	0.97	0.97	1.09
11 Einstein Method (1950- Bed-load)	0 10 20	– 28200.00 24000.00	– 29400.00 25200.00	– 5.34E+00 7.584E+00	0.82	0.83	1.53
Mean of 20 Years =		29127.27	30436.36	4.953E+00			
Standard deviation =		6942.20	6649.40	3.469E+00			

Table 2.9 Results of Total Load Equations

| | | Starting Point of Foreset Bed | Uppermost Point of Topset Bed | | Discrepancy Ratios | | |
| | | Dist. from the | Dist. from the | Height | Starting | Uppermost | Height |
Equation	Year	Dam (m)	Dam (m)	(m)	Point	Point	of Delta
1 Engelund and Hansen's Approach (1967)	0	–	–	–	1.25	1.24	0.63
	10	30600.00	31800.00	4.154E+00			
	20	27000.00	28200.00	6.008E+00			
2 Ackers and White's Formula (1973)	0	–	–	–	1.51	1.48	0.33
	10	34800.00	35400.00	2.118E+00			
	20	32400.00	33600.00	3.156E+00			
3 Bagnold's Equation (1966)	0	–	–	–	1.20	1.19	0.70
	10	30000.00	30600.00	4.318E+00			
	20	25800.00	27000.00	6.651E+00			
4 Brownlie's Method (1981)	0	–	–	–	0.11	0.16	2.02
	10	13200.00	14400.00	1.341E+01			
	20	2400.00	3600.00	1.918E+01			
5 Karim and Kennedy Approach (1990)	0	–	–	–	1.23	1.22	0.66
	10	30000.00	31200.00	4.411E+00			
	20	26400.00	27600.00	6.301E+00			
6 Yang's Gravel Formula (1984)	0	–	–	–	1.70	1.69	0.08
	10	37800.00	39600.00	3.866E−01			
	20	36600.00	38400.00	7.541E−01			
7 Yang and Lim's Formula (2003)	0	–	–	–	0.00	0.03	2.58
	10	9000.00	9600.00	1.595E+01			
	20	0.00	600.00	2.448E+01			
Mean of 20 Years =		21514.29	22714.29	9.505E+00			
Standard deviation =		14414.28	14670.09	8.808E+00			

Approach (1990), Shen and Hung's Method (1972), Engelund and Hansen's Approach (1967), Kalinske's Equation (1947) and the Molinas and Wu's Equation (2001).

2.5.5 Comparison with the sediment data of Çubuk 1 Dam Reservoir

In order to make a case study for a real reservoir-river system Çubuk 1 Dam is selected. Hydraulic and sediment data of Çubuk 1 Dam were used; the median

Table 2.10 Input table of CASE 2

Symbol	Explanation	Variable name in DELTA	Input Value	Unit
Physical Characteristics Data				
S	Initial bed slope of the River	SF	0.000175	–
d_{50}	Average Sediment Particle Diameter	D50	1.0	mm
n_M	Manning-Strickler Coefficient	CN	0.032	$m^{-1/3}s$
ρ_s	Density of Sediment Particles	ROS	2650	kg/m^3
ρ	Density of Water	ROE	1000	kg/m^3
q	Water Discharge per Unit Width	QU	1.81 and 2.5	$m^3/s/m$
Data Concerning the Modification of the Bed Profile				
$\Delta z(i)/h_i$	Maximum Relative Bed-level Change Tolerated	VARZMX	0.1	–
p	Porosity of the Sediments	POROS	0.3	–
λ	The Ratio of the Upstream/Downstream Heights of the Sediment Deposition or Erosion	HAMHAV	0.75	–
Information Concerning the Computational Domain				
x_1	x coordinate of the first station	XI	0	m
x_f	x coordinate of the last station	XF	150000	m
Δx	Step Length in x direction	DX	600	m
$\Delta V^2/2g$	Maximum Tolerated Variation of Dynamic Head	DHDYNM	0.01	m
–	Maximum Number of Subdivisions in powers of 2	NMC	7	–
Boundary Conditions				
h_1	Water Depth at the Downstream End	HI	23.5	m
Simulation Time and Printing of Results				
Δt	Time Step	DELT	*	days
–	Duration of Simulation	TFIN	7300	days
–	Step Number end of which the results are printed	NPP	**	–

*Input Value depends on the selected sediment transport equation and the deposition rate of sediment particles
**Input value depends on the choice of the time step

diameter, $d_{50} = 0.14$ mm, river slope, $S = 0.00383$, unit discharge, $q = 0.359\ m^3/s/m$, porosity, $p = 0.48$ (Kılıç, 1984). Selected sediment transport equations according to the $d_{50} = 0.14$ mm were Laursen, Rottner and Wu-Wang-Jia's equations. DELTA was employed for those selected equations for an 1835 day (5 year) simulation. The input table was shown in Table 2.15 for this run.

According to the sediment and hydraulic input values shown in Table 2.15, 5 year (1825 days) simulations of DELTA by selecting Laursen's Approach (1958), Rottner's Formula (1959) and Wu, Wang and Jia's Formula (2000) were performed. The results are presented in Figures 2.12–2.14. Detailed results of these runs are given by Pulcuoğlu (2009). Discrepancy ratios with respect to the average values of starting point distance of delta distance of uppermost point of delta and delta height showed that Rottner's

Table 2.11 Deposition percentage change due to decrease in particle size of bed-load run groups

	Equation	d_{50}			
		6.0	2.0	0.5	0.2
1	Einstein's Equation (1942)	12.30	23.04	–	–
2	DuBoys' Formula (1879)	–	0.00	66.93	178.42
3	Kalinske's Equation (1947)	17.04	12.69	–	–
4	Laursen's Approach (1958)	–	1.84	0.92	0.23
5	Rottner's Formula (1959)	–	12.78	20.31	23.04
6	Chang, Simons and Richardson's Approach (1967)	–	–	12.46	10.71
7	Wu, Wang and Jia's Formula (2000)	40.15	48.34	62.42	76.68
8	Einstein and Brown Approach (1950)	8.50	43.27	–	–
9	Yalın's Formula (1963)	0.41	0.45	0.63	–
10	Engelund and Fredsoe's Equation (1976)	18.34	52.45	–	–
11	Van Rijn's Formula (1984)	–	0.50	4.19	2.54
12	Dou's Equation (1977)	1.13	3.46	10.17	24.69
13	Wilcock and Crowe Method (2003)	10.75	30.61	55.56	–
14	Einstein Method (1950-Bed-load)	21.90	59.16	–	–

Table 2.12 Deposition percentage change due to decrease in particle size of total-load run groups

	Equation	d_{50} (mm)				
		6.0	2.0	1.0	0.5	0.2
1	Inglis-Lacey Formula (Inglis, 1968)	–	–	–	79.74	196.40
2	Engelund and Hansen's Approach (1967)	4.92	14.73	29.42	58.71	–
3	Ackers and White's Formula (1973)	–	3.46	4.52	3.87	1.66
4	Shen and Hung's Method (1972)	–	–	84.84	121.25	225.55
5	Yang's Sand Formula (1973)	–	–	37.92	37.06	57.44
6	Toffaleti Procedure (1969)	–	–	0.05	0.24	–
7	Bagnold's Equation (1966)	–	10.75	11.56	11.07	–
8	Brownlie's Method (1981)	–	89.99	208.78	–	778.56
9	Karim and Kennedy Approach (1990)	2.14	12.81	–	–	–
10	Bishop, Simons and Richardson Method (1965)	–	–	100.50	41.80	–
11	Yang's Gravel Formula (1984)	0.93	0.38	–	–	–
12	Yang and Lim's Formula (2003)	–	145.58	221.62	–	–
13	Molinas and Wu's Equation (2001)	–	–	55.42	71.43	–

Formula gives the closest values to the mean values. However, the sediment data of Kılıç (1984) indicates that delta formation height for 50 years is approximately 5.5–6.0 meters which spreads over 3.5 km long distance measured approximately 3.5 km to 6.0 km from the dam (Kılıç, 1984). This region where river meets reservoir referred as muddy region (see Figure 2.11). It is obvious that simulations made by employing Rottner's and Wu-Wang-Jia's Formula's over estimated the delta formation in terms of both placement and height of delta. Disregarding Rottner's and Wu-Wang-Jia's formulas, the Laursen Approach is employed for a 50 year simulation. The results of Laursen's

Table 2.13 Deposition percentages with respect to the change of unit discharge of bed-load equations

	Equation	Deposition Percentage q (m³/s/m)	
		1.81	2.50
1	Einstein's Equation (1942)	2.20	2.85
2	DuBoys' Formula (1879)	0.00	0.00
3	Kalinske's Equation (1947)	1.54	1.76
4	Laursen's Approach (1958	0.05	0.10
5	Rottner's Formula (1959)	1.19	1.97
6	Chang et al. Approach (1967)	0.04	0.06
7	Wu, Wang and Jia (2000)	3.15	5.05
8	Einstein and Brown Approach (1950)	2.41	4.21
9	Yalin's Formula (1963)	0.05	0.07
10	Engelund and Fredsoe's Equation (1976)	4.55	6.37
11	Van Rijn's Formula (1984)	0.03	0.09
12	Dou's Equation (1977)	0.23	0.40
13	Wilcock and Crowe Method (2003)	2.24	3.38
14	Einstein Method (1950-Bed-load)	5.41	7.26

Table 2.14 Deposition percentage change due to the river slope and discharge changes

	Equation	Deposition Percentage			
		S		q (m³/s)	
		0.000450	0.000175	1.81	2.5
1	Engelund and Hansen's Approach (1967)	29.42	1.74	1.03	1.74
2	Ackers and White's Formula (1973)	4.52	0.86	0.46	0.86
3	Shen and Hung's Method (1972)	84.84	1.76	0.80	1.76
4	Yang's Sand Formula (1973)	37.92	2.68	1.57	2.68
5	Toffaleti Procedure (1969)	0.05	0.00	0.00	0.00
6	Bagnold's Equation (1966)	11.56	1.17	0.80	1.17
7	Brownlie's Method (1981)	208.78	11.31	5.71	11.31
8	Bishop et al. Method (1965)	100.50	3.67	3.61	3.67
9	Yang and Lim's Formula (2003)	221.62	16.68	11.21	16.68
10	Molinas and Wu's Equation (2001)	55.42	5.05	1.28	5.05

Approach for a 50 year simulation are approximate to Kılıç's (1984) observations in terms of the placement and height of delta deposition (see Figure 2.15). However, it should be noted that cross-sectional view of Çubuk 1 reservoir (Figure 2.11) includes the accumulation of fine sediments which is not in the scope of this study since they are below the sand size range. Even though it is quite difficult to distinguish in Figure 2.11 the regions of delta deposit, i.e., deposition of relatively coarser particles, and accumulated finer particles a comparison can be made with respect to the muddy region (delta deposit region) thickness and extend as stated by Kılıç (1984).

Table 2.15 Input table of Run related to the Çubuk I Dam

Symbol	Explanation	Variable Name in DELTA	Input value	Unit
Physical Characteristics Data				
S	Initial bed slope of the River	SF	0.00383	–
d_{50}	Average Sediment Particle Diameter	D50	0.1	mm
n_M	Manning-Strickler Coefficient	CN	0.032	$m^{-1/3}s$
ρ_s	Density of Sediment Particles	ROS	2350	kg/m^3
ρ	Density of Water	ROE	1000	kg/m^3
q	Water Discharge per Unit Width	QU	0.359	$m^3/s/m$
Data Concerning the Modification of the Bed Profile				
$\Delta z(i)/h_i$	Maximum Relative Bed-level Change Tolerated	VARZMX	0.5	–
p	Porosity of the Sediments	POROS	0.48	–
λ	The Ratio of the Upstream/Downstream Heights of the Sediment Deposition or Erosion	HAMHAV	0.75	–
Information Concerning the Computational Domain				
x_I	x coordinate of the first station	XI	0	m
x_f	x coordinate of the last station	XF	10000	m
Δx	Step Length in x direction	DX	50	m
$\Delta V^2/2g$	Maximum Tolerated Variation of Dynamic Head	DHDYNM	0.05	m
–	Maximum Number of Subdivisions in powers of 2	NMC	7	–
Boundary Conditions				
h_I	Water Depth at the Downstream End	HI	23	m
Simulation Time and Printing of Results				
Δt	Time Step	DELT	0.1	days
–	Duration of Simulation	TFIN	1825*	days
–	Step Number end of which the results are printed	NPP	3650	–

*Only Run with Laursen Equation was made for both 5 years (1825 days) and 50 Years (18250 days)

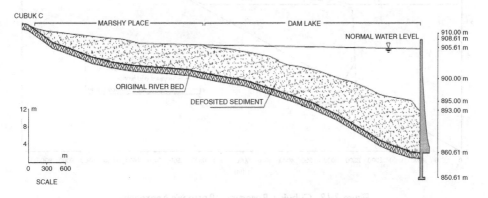

Figure 2.11 Cubuk Dam (Adopted from Kılınç, 1984)

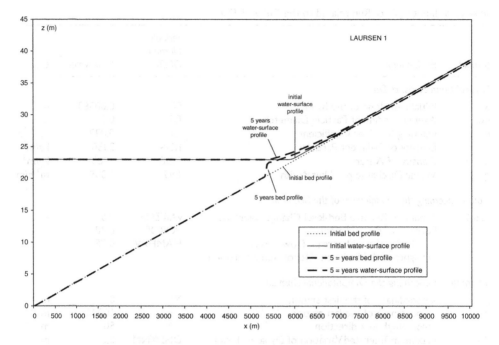

Figure 2.12 Cubuk I Reservoir (Loursen's Approach)

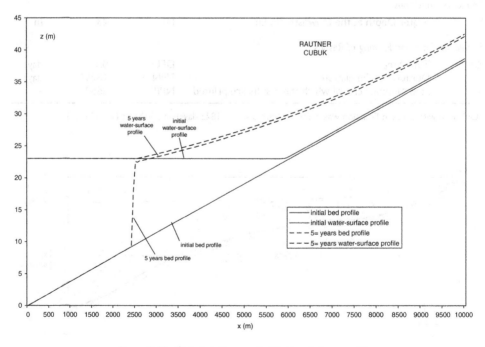

Figure 2.13 Cubuk I Reservoir (Rottner's Approach)

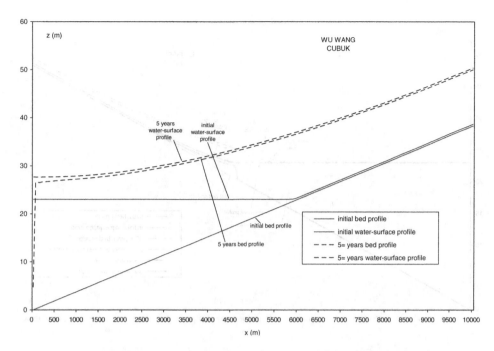

Figure 2.14 Cubuk I Reservoir (Wu et al. Approach)

Although reservoir sedimentation includes the accumulation of relatively coarser and finer sediment particles in reservoirs, DELTA program which is developed by Graf and Altınakar (1998) is designed for the estimation of deposition of coarser particles at the entrance of reservoirs. Because of this, an extension of the DELTA program is made by adding the codes of suitable sediment transport equations that are derived especially for sand and gravel size ranges. As a consequence, calculated deposition percentages for various simulation periods do not include the siltation amount of finer particles below sand size range.

DELTA is based on some rough assumptions that seem to be distant from real river-reservoir systems such as constant discharge, prismatic channel width (even including reservoir up to dam section), constant Manning's roughness coefficient and constant river slope. Even though the effect of these parameters on sediment transport is obvious, the simplification by those mentioned rough assumptions gives convenience to estimate the depositions by comparing sediment transport equations that are based on various similar assumptions.

The selection of space interval and time-step length is important in order to get reasonable values on delta formation. To get a good representation of delta formation the space interval must be small enough. On the other hand, too short a time-step increases computation time and may lead to computational errors due to very small deposition values close to the precision of the floating point numbers in the computer. For instance, some of runs performed for simulation of Çubuk 1 reservoir led to computational errors. Since the Çubuk 1 reservoir-river system has a relatively steep

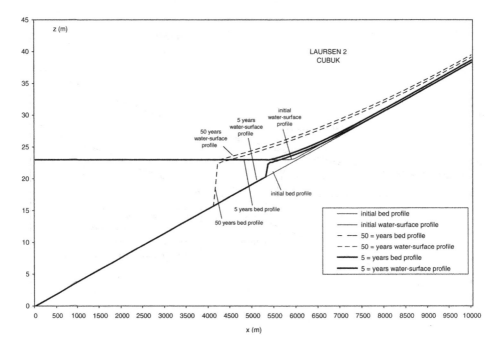

Figure 2.15 Cubuk I Reservoir (Laursen's Approach)

slope and less water discharge, calculated normal depth of the river is small. At the point where river meets reservoir in the Çubuk 1 reservoir-river system, the abrupt change in the flow depth leads to rapid decrease in sediment carrying capacity causing abnormal increase in aggradation rate. A very small time step is chosen, in order to avoid from this effect, however this time simulation process stopped with abnormal end messages due to computational errors. This is the main reason why only three of the available equations could have been used for a simulation of the Çubuk 1 reservoir even though there are some equations among the 32 that are compatible with 0.14 mm median particle size.

Chapter 3

Techniques for preventation of sediment deposition

The main reason for losing water storage capacities is siltation in the basin area. There are many different techniques throughout the world to overcome this siltation problem. Some of these techniques deal with inflowing sediment. On the other hand, some of them try to remove siltation. However, throughout the world the problematic regions do not use the same techniques because every basin has different characteristics; geological, geographical and climatic, etc. For example, the dredging technique is mostly used in semi-arid regions like China. On the other hand, in the north European countries, in rainy regions, flushing techniques are used.

Nonetheless, in this study the basin water storage capacity management techniques used throughout the world are categorized as follows:

1   Preventing Sediment Inflow into the reservoir,

    –   Watershed Management
    –   Upstream Check Structures (Debris Dams)
    –   Reservoir Bypass

2   Sustainable Management of the Reservoir,

    –   Evacuation of Sediments from Reservoir
        •   Flushing
        •   Sluicing
        •   Density Current Venting
        •   Mechanical Removal
            •   Dredging
            •   Hydrosuction Removal System (HSRS)
            •   Trucking
    –   Management of Reservoir
        •   Operation Rules
        •   Tactical Dredging

3   Lost Storage Replacement Techniques or Decommissioning of Dam (Retirement of dam).

    –   Raising Dam Height
    –   Build New Dam
    –   Decommissioning

3.1 PREVENTING SEDIMENT INFLOW

The sediment silted in a reservoir causes several major problems. If the sediments coming from the upper reaches of the river could be stopped or diverted before reaching the dam body, then the majority of problems will decrease. There are some methods for preventing sediment inflow explained in the following sections.

3.1.1 Watershed management

Watershed management is a method which is used to reduce the reservoir siltation coming from the upstream basin of the reservoir by using some techniques, such as forestation, prevention of erosion by vegetation and tillage management, sediment trap and change in land usage. As a matter of fact, watershed management aims to conserve soil and consequently conserve water. In order to achieve this aim, techniques should be combined efficiently.

According to White, the storage losses in China and India are because of low forest covers, which are 16.5% and 23% respectively. As shown in Table 3.1, China and India are losing their storage capacity, approximately 2.3% and 0.46%. On the other hand, the storage losses of Japan and Southeast Asia are lower than China and India as a result of having high forest cover (White, 2000). Hence, in order to control the amount of sediment entering a reservoir, it is recommended that the soil surrounding the reservoir should be controlled, in other words there should be watershed management.

Although, watershed management is one of the recommended reservoir sedimentation prevention techniques in the literature, there are some opposite research results about its efficiency. For example, an extensive watershed management project was executed in Mangla watershed, in Pakistan, before the dam construction.

Mangla Dam is a multipurpose, 112 m high, earth-rockfill dam on Jhelum River in Pakistan with a crest level of 376.1 m (Mahmood, 1987). Its catchment area is 33,333 km^2 and total storage capacity is 9.47 km^3. The schematic catchments area of Mangla Dam is shown in Figure 3.1. From the data shown in Figure 3.1, the highest sediment concentration is coming from Kanshi River.

The main object of the Mangla watershed management project was reducing the sediment load at Mangla. Project was started at 1959 and it contains a large number of structural and non-structural measures. The whole project duration was 30 years such that the observation phase took 7 years (1959–1966) and the operation phase took 23 years. For the purpose of evaluation of sediment loads, gauging stations were used. Data coming from the gauging station shows that no discernible difference in sediment loads is noted over a period of 4–14 years (Mahmood, 1987). As a consequence, the Mangla watershed management project effected the local environment and productivity positively. However, for the purposes of decreasing sediment load in Mangla watershed, its contribution is insignificant.

Actually, especially for semi-arid regions, the reservoir shoreline management should be considered alongside river basin management. In the shoreline there would be high erosion rates. However, it is not feasible to protect the whole shoreline against erosion because of its long length, except in localized specific areas where high value property or structures are threatened (Morris and Fan, 1997). The use of riprap, sheet

Table 3.1   Storage loss rates in different countries (Liu, Liu and Ashida, 2002)

Location	Annual Sedimentation Rate (%)	Total Sedimentation Rate (%)	Source
China	2.30	14.2	Hu, 1995
India	0.46	9.6	White, 2001
Japan	0.15	8.8	ANRE, 1984–2000
Southeast Asia	0.30	8.0	White, 2001
South Africa	0.34	11.4	White, 2001
Turkey	**1.50**	**59.7**	**White, 2001**
UK	0.10	–	White, 2001
USA	0.22	3.9	Morris and Fan, 1997
World	1.00	11.8	White, 2001

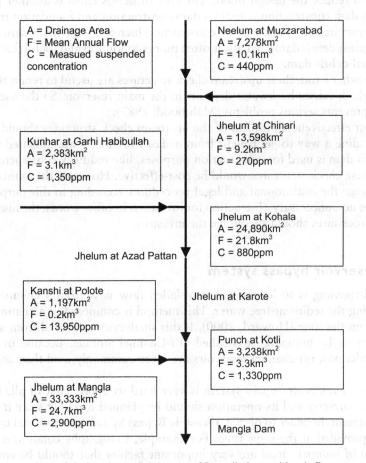

Figure 3.1   Schematic Catchment of River Jhelum at Mangla Dam

piling and reduction of boat speeds to minimize wake, which is the path left by boat in water, are some shoreline erosion prevention practices.

3.1.2 Upstream check structures (Debris dams)

Debris dams are relatively smaller in size than the main dam provided by spillway structures. Their function is to stop the major part of the incoming sediment coming from tributaries to the reservoir of a dam which will be used to supply either power or water supply. It can definitely extend the life of the dam but in return the debris dam itself will require a reservoir sediment management program.

From an economical viability point of view, the short lifetime and relatively high cost of debris dams are the problem. Sediment concentration of tributary conversely effects the lifetime of a debris dam. This means, if sediment concentration in the tributary is large, the lifetime of the debris dam is short. In addition, debris dams are not active to reduce the design flood. The cost of debris dams is another problem, since debris dam construction is again a dam construction and foundation treatment, auxiliary structures etc. must also be constructed. Increasing the main dam capacity, weighed against debris dam capacity, is often more economical and useful than to build an additional debris dam.

On the other hand, these upstream check structures are useful to retain the coarse material, which causes backwater deposits in the main reservoir. So the use of these structures prevents serious problems (Mahmood, 1987).

The cost effectiveness problem of the upstream check structures should be overcome by finding a way to get benefit from it. For example, if accumulated sediment in the debris dam is used for construction purposes, like roadway construction material, etc., these check structures would be cost-effective. However, the authorizations should arrange the institutional and legal procedures according to this purpose, that there will be no unnecessary obstruction for investors. In other words, the institutional and legal procedures should encourage the investors.

3.1.3 Reservoir bypass system

Reservoir bypassing is to let the sediment-laden flow pass from a channel, meanwhile keeping the sediment-free water. This method is composed of a channel within a reservoir on the river (Howard, 2000). If this small reservoir is fed from a river by main gravity or by pumping, it is called Off-Channel Storage. Because of environmental restrictions, off-channel reservoirs are more commonly used than on-channel reservoirs.

Actually, a reservoir bypass system is very hard to apply. First of all, it should be designed correctly and its operation should be planned carefully since it is a very expensive system. In order to obtain a feasible Bypass System some special conditions should be provided at the same time. For example, topography conditions and size distribution of sediment load are very important factors that should be considered. Sediment excluders (sand traps) can be used in arid areas. Bypassing sediment-laden water from a channel is not acceptable for arid areas, where the need for water is high. However, it should not be expected that sediment excluders will remove significant quantities of silts and clay; actually of the sand load, the excluders can optimally remove only about half of the load with one-tenth of the flow (Mahmood, 1987).

It should be noted that water during the flood is diverted by the bypass system, there is no need for a large-capacity spillway at the main dam.

Some reservoirs that use the bypass system effectively are (Howard, 2000): Gmünd Reservoir in Austria, Tedzen Reservoir in Turkmenistan, Amsteg Reservoir and Palagnedra Reservoirs in Switzerland, where several reservoirs have a bypass system, and Syiya, Yanshuigon and Lushuihe Reservoirs in China.

3.2 SUSTAINABLE MANAGEMENT OF THE DAMS

3.2.1 Evacuation of sediments from reservoir

3.2.1.1 Flushing

Definition of flushing

Flushing is a sediment removal technique where deposited sediment is scoured from the reservoir, by increasing the flow velocity, and then transported through low level outlets. Flushing can be operated in two ways: by lowering the water level or without lowering the water level, called "pressure flushing" and "empty (free-flow) flushing", respectively. Pressure flushing is to release water through the bottom outlets by keeping the reservoir water level high. Nonetheless, it is not a commonly used technique and it is less effective than empty (free-flow) flushing. On the other hand, empty (free-flow) flushing is releasing water by emptying the reservoir and also to route inflowing water from upstream by providing riverine conditions. There are two types of empty flushing: flood season and non-flood season. Both of them are successful in practice, flood season flushing is more effective since it provides larger discharges to route the sediment (Morris and Fan, 1997).

Flushing is not used widely because of some restrictions about its effectiveness. For example, flushing is generally effective in narrow dams. In addition, a considerable amount of water should be passed through the reservoir for drastic flushing operation. Also, the most important restriction is that, the reservoir is required to be empty for drawdown flushing (the most effective technique). This is the most limiting condition, because for hydropower dams being empty is not acceptable from an energy point of view. Moreover, the water released in flushing has a very high sediment concentration, much higher than in natural riverine conditions. This extreme concentration can cause unacceptable conditions for downstream.

Criteria for flushing operation

Basson and Rooseboom (1997) prepared a diagram (Figure 3.2), where 177 dam cases had been reviewed for the selection of reservoir operation. This diagram enables us to make a preliminary judgement about whether flushing is an effective technique or not. This judgement is made by calculating K_w and K_t which are the ratios of storage to mean annual river run-off and storage to mean annual sediment yield respectively, and then see in which zone of the diagram the controlled reservoir takes place.

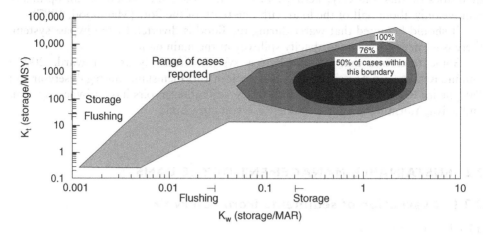

Figure 3.2 Basson's diagram on reservoir operation modes. Modified from Basson (1997)

The empirical indices in Basson's diagram are:

$$K_w : C_0/MAR \qquad (3.1)$$

$$K_t : C_0/MSY \qquad (3.2)$$

in which C_0 is reservoir capacity, MAR is mean annual river run-off and MSY is mean annual sediment yield.

The right upper part of the diagram, where K_w is larger than 0.2, is denser than the other parts since most of the reservoirs in the world have been designed for 100 or more year's sediment accumulation and these reservoirs have not enough water for flushing operation or reservoir drawdown. Because of not having excess water for flushing, density current venting can be practiced at these reservoirs and dredging can be used to recover lost storage capacity (Basson, 2004). In regions where K_w value is variable from 0.03 to 0.2, seasonal flushing is suggested. Furthermore, in case K_w is smaller than 0.03 sediment sluicing and flushing should be carried out during floods and through large bottom outlets, preferably with free conditions especially in semi-arid regions (Basson, 2004). On the other hand, according to the ratio K_t, lower part of the diagram where $30 < K_t < 100$ excess water is available and flushing is efficient.

Flushing operation selection case study: Çubuk Dam-I

As a case study, Çubuk Dam I is analyzed where the ratios stand in Basson's Diagram.

$$C_0 = 7.1 \, mm^3$$

$$MAR = 65.5 \, mm^3$$

$$MSY = 60,000 \, m$$

$$K_w = 7.1/65 = 0.1092$$

$$K_t = 7,100,000/60,000 = 118$$

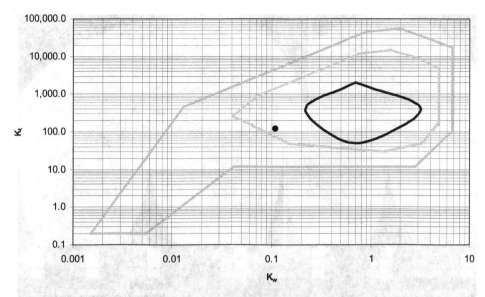

Figure 3.3 Çubuk Dam-I in Basson's Diagram

Çubuk Dam-I is in the zone of 75%, K_t value is not appropriate for flushing but K_w is between 0.03 and 0.2. Therefore, seasonal flushing can be suggested according to Basson's empirical table.

If appropriate sediment management studies were carried out during the planning stage, it would have been possible to apply seasonal flushing. That is, Çubuk Dam-I wouldn't have been out of function as it is today.

Factors affecting flushing viability and efficiency

Hydraulic requirements, available water quantity, sediment mobility in reservoir and special conditions of site are factors that affect the flushing efficiency (Howard, 2000). These factors are discussed below:

- Hydraulic requirements: for flushing operation hydraulic conditions in the reservoir should be the same as river conditions. This river condition maintained by sufficient bypass capacity, at least twice of mean annual run-off as flushing discharge and at least 10% of mean annual run-off as flushing volume (Howard, 2000).
- For transporting enough sediment from the reservoir those having smaller capacity compared with annual run-off, are preferred. Come to that, demand water should always be balanced (Howard, 2000).
- In order to determine the required water for flushing, sediment sizes and sediment mobility should be defined well. It is known that fine sand and coarse silt are the most successful sizes for flushing (Howard, 2000).

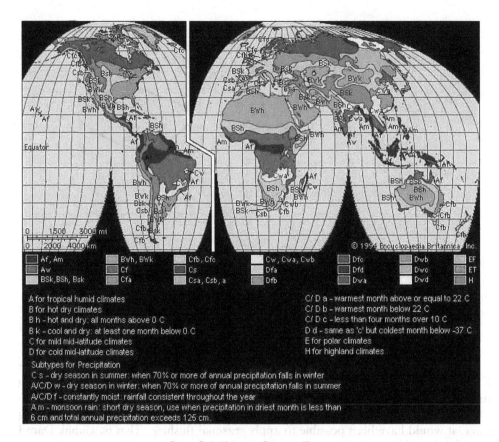

Figure 3.4 Koppen Climatic Zones

- The width and shape of the reservoir are the important site specific features for flushing. Because long and narrow reservoirs are more suitable for flushing operation (Howard, 2000).

In practice, geographical factors also give signs to help decide if the area is suitable for flushing or not. These factors are:

- Erosion rate.
- Sediment Yield.
- Climatic Zones: According to the Koppen classification of climatic zones, the best flushing efficiency is gained in "Tropical Wet and Dry Region" and it can be also said that "Tropical Wet Region" is also suitable (Figure 3.4). These regions are:

  • Parts of Central America extending into Brazil in South America.
  • A region of Central Africa from the Ivory Coast in the west to Sudan in the east would be suitable.

- Parts of central Asia including Pakistan, India, Nepal, China, Vietnam and Thailand (Howard, 2000).

As mentioned in the above paragraph, Turkey does not have a place in the list of best flushing regions according to climatic zone. Indeed, in the map of climatic zones the middle Anatolia is seen as the defined region and it is suitable for flushing in terms of climatic conditions.

3.2.1.2   Sluicing

Sediment sluicing is defined as a worldwide operational design where, in most cases, the reservoir is drawn down in the flood season and then sediment carrying inflow is directly passed through the reservoir (so that sediment has no chance to settle down). After the flood season, clear water will be stored and the reservoir will be raised for the next season usage.

The reason why this method is applied during the flood season is to get sufficient sediment to be transported. During the rising water level of a flood the out-flowing sediment discharge is always smaller than that of the inflow, due to the backwater effect causing a substantial decrease in the velocity. Oppositely, during the lowering of the water level, the out-flowing sediment discharge is greater than the inflow because there is no backwater effect and there is erosion in the reservoir (Fan, 1985). Therefore, sluicing in flood season is efficient from a hydraulics point of view.

One of the advantages of sluicing over flushing is that the sediment problems of downstream reaches regarding high concentration flow will be minimized by using this method. This is because the transport capacity of downstream flood discharge is greater than the transport capacity of reservoir flood discharge (Fan and Morris, 1992b).

There are also other advantages of sluicing over other evacuation methods. For example, sediment concentration of released water to downstream is lower in the sluicing operation than in the flushing operation (Morris and Fan, 1997). Furthermore, consolidated cohesive sediment movement in significant amounts is impossible by using the flushing operation (Basson and Rooseboom, 1997). Also, the velocity needed to move the eroded sediments is much higher than the velocity to keep the sediments suspended (Basson and Rooseboom, 1997). In other words, sluicing is preferable since it maintains incoming sediment in suspension.

Excess run-off availability, grain size of sediments and reservoir morphology are the main factors that affect sluicing efficiency. Worldwide, sluicing and flushing are used together.

Sluicing usage in the World

The Aswan Dam was built during 1898–1902 on the river Nile, to provide summer irrigation supplies to Middle Egypt. It was raised twice in 1912 and 1933 (Mahmood, 1987). The reservoir storage capacity is 5.6 km^3 and the annual run-off was estimated at 84 km^3. The dam had 180 sluice gates in four groups. The sluices had a 2,240 m^2 cross sectional area with 6,000 m^3 flood discharge capacity during normal flood level or more than twice flow rate during high flood. The gates are kept open during the flood months of July, August and September. Hence, from October, the reservoir is

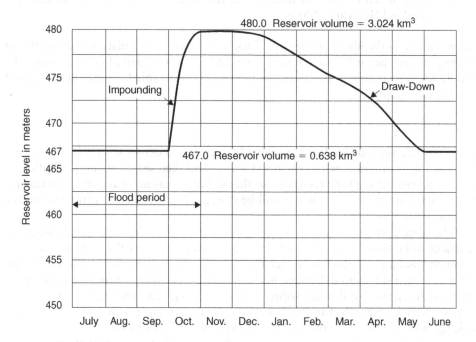

Figure 3.5 Design Operating Program for Roseires Dam: Median Inflow and full use of storage (after Schmidt, 1983; Mahmood, 1987)

filled with clear water up to 121 m and kept constant to provide sufficient irrigation requirements. According to this operation, surveys show that the sedimentation is not significantly affecting the reservoir.

Another example is Roseires Dam on the Blue Nile in Sudan. The reservoir storage capacity is 7.4 km^3 and the annual run-off was estimated at 50 km^3. The proposed operation program is shown in Figure 3.5 (Mahmood, 1987).

According to the operation program, it was planned that in the flood period – July, August, and September – the reservoir level will be maintained at 467 m. After the flood period, during October, the reservoir will be filled up to 480 m. Then up to the end of May drawdown is done to the elevation 467 m. However, the efficiency of this sluicing operation is not as effective as expected. Reasons for that difference can be seen from the comparison depicted in Table 3.2 (Mahmood, 1987).

Reservoir width at the maximum height is nearly five times greater in Roseires Dam. Because of this, the sediments are deposited on the overbank area and the sluicing cannot affect this area. Subsequently, if a sluicing operation is designed, reservoir morphology should be taken into consideration.

3.2.1.3   Density current venting

The literal meaning of density current is the movement of two fluids, with a similar state, towards each other because of different densities. Venting density currents means to route the sediment-laden flow through the stored water in the reservoir. Then the

Table 3.2 Comparison of Aswan and Roseires Dams (Mahmood, 1987)

COMPARISON FOR ASWAN AND ROSEIRES DAMS

	Old Aswan	Roseires
River Bed Level, m	87.5	435.5
Conservation Pool Level, m	121	480
Height of Conservation Pool above River Bed, H, m	33.5	44.5
Mean Annual Flow, km³	84.0	50.0
Capacity at Conservation Pool, km³	5.6	3.0
Capacity: Inflow	0.067	0.060
Annual Sediment Load, mm³	80.0	86.6
Dam Length, L, km	2.14	13.5
L/H	63.9	303.4
Measured Trap Efficiency, percent	0.0	46.0

sediment-laden flow will get to the downstream. From Figure 3.6, the movement of the density current through the reservoir and vented through a low-level outlet could be observed.

In order to have successful density current venting, the incoming sediment-laden flow would have enough velocity and fine particles to form turbid flow. If these favourable conditions exist, current should be able to reach the dam. Then the bottom outlet will be opened causing the current be vented through. The timing of density current venting is very important. Chen and Zhao (1992) have pointed out that the timing of gate opening and closing is very crucial. If sluice operation is too late or sluice gates openings are too small, these cause only small amounts of sediment discharge. If the gate is opened too early or the opening is too large, not only the loss of valuable water occurs, but also a strong velocity field of clear water will be formed in front of the outlet and prevent the turbid flow from entering the outlet (Chen and Zhao, 1992).

Density Current Venting efficiency depends on factors such as reservoir topography, thermal and salinity related stratifications, conditions of incoming flow, sediment characteristics and outlet facilities. It is obviously seen that, because of all these variables, there is an uncertainty in the flow path of density current. Providing multi-level multiple outlets is a way to overcome this problem (Mahmood, 1987).

Another advantage of density current venting is that there is no need to decrease the reservoir water level, unlike flushing or sluicing.

Density current venting method usage in the World

The usage of this method is not common on a global scale, but there are some cases of usage. For example, Bajiazui Dam in China is a very successful enlargement of this operation method. The density current venting provides an average release ratio of 46% (Howard, 2000). On the other hand, in Sefid Rud Reservoir in Iran the venting of density current is not very effective compared to Bajiazui Dam: 6 million tons of sediment was released by density current venting but the coming sediment

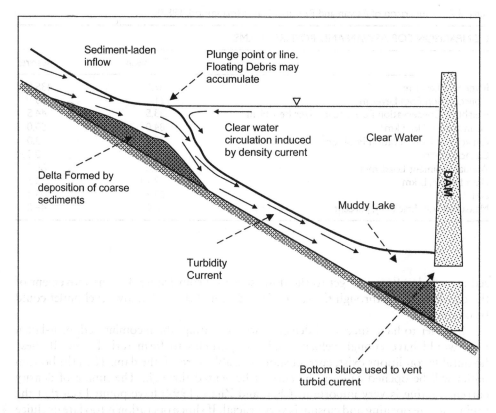

Figure 3.6 Schematic Diagram of the passage of a turbid density current through a reservoir and being vented through a low level outlet

is approximately between 40 million to 50 million tons (Parhami, 1986; Amini and Fouland, 1985; Howard, 2000). Therefore, the original storage of dam, 1.8×10^9 m^3 in 1962 has decreased to 1×10^9 m^3 in 1981 (Howard, 2000).

3.2.1.4   Mechanical removal

Evacuating sediments from the reservoir is not only done hydraulically but also done mechanically. The mechanical removal techniques are classified as dredging, hydro-suction removal system (siphoning), and trucking (dry excavation). The techniques are explained below in more detail.

Dredging

Dredging is picking up the accumulated sediment from the reservoir or lake bed and then transporting it to another area. There are several types of dredging equipment in the conventional usage.

Dredging systems can be classified as hydraulic and mechanical. In the hydraulic systems, the deposited sediment is lifted with water and then this sediment-water slurry

Figure 3.7 Hydraulic dredging equipment

formation transported from the output point to the point of placement (see Figure 3.7). Hydraulic dredging is more widely used than mechanical dredging. Hydraulic dredging advantages are:

− Low unit cost of sediment removal.
− Production rates are high.
− Working ability in a reservoir that does not interfere with the impoundment process.
− Effective both fine and large materials removal, but larger materials removal have higher cost.

The disadvantage of hydraulic dredging is the difficulty to bulk the fine sediments to the point of placement and the need for dewatering operation for lifted sediment-water slurry.

Mechanical systems, on the other hand, use buckets in order to dig the reservoir bed and then pick up the sediment. Figure 3.8 and 3.9 are examples of mechanical dredging equipment.

The sediments, dredged by using mechanical dredging equipment, have low water entrainment. Therefore, for arid areas, mechanical dredging is more advantageous

Figure 3.8 Hydraulic dredging disposal example

Figure 3.9 Mechanical Dredging Equipment

than hydraulic dredging. However, mechanical dredging production rate is lower than hydraulic dredging.

Dredging method is suitable for medium and small size reservoirs, which do not have enough water for flushing. Dredging is used for the removal of coarse sediment. The most important difficulty with dredging is to find a suitable area for damping the removed sediments. Therefore the cost of disposal land is an important item in the calculation of dredging cost. A compilation of dredging cost has been carried out by Basson and Rooseboom (1997).

Dredging is the most commonly used technique in China and Japan. Especially in Northwest China, because the dredging cost is relatively cheap; dredging has been used for more than 10 reservoirs since 1975 (Wang, 1996).

As listed in Table 3.3, the unit cost of dredging and cost of siphoning system are 0.21 RMB/m^3 and 103.7 × 10^3 RMB (Chinese Yuan), respectively, for Xiaohuasha reservoir. This means that, since dredged sediment quantity is 406 × 10^3 m^3, the total cost of dredging is 85.26 × 10^3 RMB. Obviously cost of dredging is relatively cheaper than the cost of siphoning.

In addition, they overcome the disposal area problem by transporting dredged sediment through irrigation canals to recharge the topsoil of cropland (Wang, 1996).

Hydrosuction Removal System (HSRS)

Hydrosuction Removal System (HSRS) is simply a siphon and airlift system, which uses the potential energy stored by the hydraulic head at the dam, removes the sediments through a floating or submerged pipeline to an outlet. This system is known as the Hydrosuction Removal System (HSRS) and Sediment Evacuation Pipeline System (SEPS) in the USA, and Geolidro System in Italy (Liu, Ashida and Hindley, 1994; Hotchkiss and Huang, 1995).

The system is composed of a barge, a pipeline and appurtenant valves to control flow. The barge is used to control the flow upstream and downstream of the pipeline, and also move the upstream end of the pipe in order to provide movement of the suction head of the pipe. A sketch of HSRS system is shown in Figure 3.10. As depicted in the sketch the silted sediments removed through pipeline to downstream end of the reservoir (Figure 3.11).

HSRS has some advantages over dredging. For example, since potential energy stored by the hydraulic head is used as a driver there is no need for any equipment to produce energy. Therefore, the operating costs are substantially lower than those of traditional dredging. In addition, there is no need to find a suitable disposal site with HSRS, since sediment is moved to the downstream end of the reservoir; an environment friendly operation.

On the other hand, the major disadvantage of HSRS is that the system can be used in relatively short reservoirs, no longer than approximately 3 km, and also dependent on the elevation of the dam and reservoir (Palmieri et al., 2003).

HSRS is one of the less frequently used techniques for sediment removal (Hotchkiss and Huang, 1995). However, in China this technique is used effectively. The first usage of a siphon system for sediment removal is applied in the Tianjiawan Reservoir. The system consists of a barge and a floating pipeline of 229 m in length and 550 mm in diameter connected to the dam outlet (Liu, Liu and Ashida, 2002).

This technique is used in the case of a lack of sufficient amount of water for flushing. The released water and sediment was used for irrigation.

Trucking (dry excavation)

Trucking (also known as dry excavation) is excavation of the accumulated sediment from reservoir like dredging, but it requires drawdown of reservoir. The excavated

Table 3.3 Sediment removal practices in China by siphon dredging system

Reservoir	Tianjiavan	Xiahuasha	Youhe	Xihe	Taoshupo	Beichaji	Xihe
Damsite (city, province)	Yuci, Shanxi	Huaxian, Shaanxi	Weinan, Shaanxi	Lintong, Shaanxi	Fengxiang, Shaanxi	Nijing, Gansu	Lintong, Shaanxi
Dam Height (m)	29.5	33	32	35	32.5	15	39.8
Original Storage Capacity (10^6 m^3)	9.43	1.77	24.5	3.94	1.54	2.75	6.74
Annual run-off (10^6 m^3)	3.95	2.5	33.6	4.6	1.31	1.09	3.69
Annual sediment load (10^3 ton)	320	50	650	450	73	163	440
Annual sediment concentration (kg/m^3)	82	20	19.3	87.5	55.6	150	–
Completion year	1960	1959	1959	1969	1959	1972	1971
Survey year	1978	1978	1978	1978	1982	1978	1978
Operation period (year)	18	19	19	9	23	6	7
Total deposition (10^3 ton)	4,000	525	8,985	3,100	782	528	5,780
Annual deposition (10^3 ton)	220	276	473	344	34	88	340
Discharging structure	Outlet conduit	Pivot Gate	Draw-off tower	Flexible pipe	Outlet conduit	Bottom outlet	Outlet conduit
Discharging capacity (m^3/s)	2.5	5	5.5	2	–	10	2
Starting year of dredging	1975	1976	1976	1977	1978	1977	1978
Gross work head (m)	17.4	–	–	–	–	–	–
Effective work head (m)	5.5–8.8	0–20.8	9–14.0	5–10.0	6–8.0	6–14.0	–
Diameter of pipe (m)	0.55	0.3	0.72	0.3	0.25	0.23–0.5	0.18
Length of pipe (m)	229	83–350	–	–	–	–	–
Pipeline discharge (m^3/s)	1.2	0.3	0.72	0.3	0.25	0.23–0.5	0.18
Suction head type	Aspirator	Aspirator	Cutter	Cutter	Aspirator	Aspirator	Cutter
Water depth of reservoir	3	12.8–2.7	6–8.0	2–8.0	2–6.0	15–5.0	–
Mean sediment concentration (kg/m^3)	190	136–168	–	–	87.7	484	3
Max sediment concentration (kg/m^3)	480	720	581.5	1,080	50.5	1,143	–
Use of release sediment	Irrigation	Irrigation	Irrigation	Irrigation	Irrigation	Irrigation	Irrigation
Grain size at outlet, d_{50} (mm)	0.007	0.024–0.044	0.015–0.029	–	–	0.005–0.096	–
Dredging Period	1975–77	1976–86	1978–86	–	–	1078–86	–
Dredging hours	695	4619	4203	–	–	4353	–
Dredged Sediment (10^3 m^3)	320	406	325	–	–	213	–
Hourly dredged sediment (m^3/hr)	460	88	77	–	–	49	–
Cost of siphon dredging system (10^3 RMB)	–	103.7	99.8	–	–	32.2	–
Dredging unit cost (RMB/m^3)	0.045	0.21	0.16	–	–	0.22	–

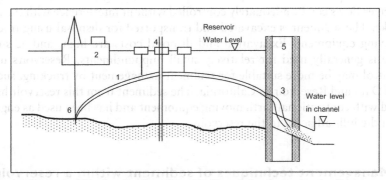

1 = Submerged Pipeline, 2 = Operation ship (Barge), 3 = Connector, 4 = Pontoons,

5 = Outlet Equipment, 6 = Suction Head, 7 = Deposit Surface

Figure 3.10 Siphon Dredging System at Tianjiawan reservoir (after Zhang and Xie, 1993)

Figure 3.11 Silted sediments removed from pipeline

sediment is transported to a suitable disposal area by using traditional earth moving equipment. Cost of drawdown, transportation cost and suitable area cost makes the method very expensive. That is why the trucking method is not widely used. Indeed, dry excavation methods are generally more expensive than dredging. They are more often used along the upper reaches of reservoirs (Howard, 2000).

Trucking requires the lowering of the reservoir during the dry season when the reduced river flows can be adequately controlled without interference with the excavation works. The sediment is excavated and transported for disposal using traditional earth moving equipment. Excavation and disposal costs are high, and as such this technique is generally used for relatively small impoundments. Reservoirs used for flood control may be more suitable for sediment management by trucking, such as at Cogswell Dam and Reservoir in California. The sediment from this reservoir has been excavated with conventional earth moving equipment and has been used as engineered landfill in the hills adjacent to the reservoir.

3.2.2 Management techniques of sediment within a reservoir

–   *Operating Rules*
    The reservoir operating rule can affect where the sediment deposition occurs. For example, during flood season if the reservoir water level is high then the sediment is mostly deposited in the upper reaches of the reservoir. On the contrary, during flood season if the reservoir is drawn down then the sediments tend to deposit in a dead storage zone of the reservoir.
–   *Tactical Dredging*
    There is one more dredging type: Tactical Dredging, which is used for local sediment removal. For example, for dams built for hydropower generation it is important to keep the vicinity of the outlets clear in order to prevent blockage of the outlets. Blocked outlets will cause energy production to stop. Furthermore, the mechanical equipment, like turbines, will be damaged because of sediment. Thus, the useful life of the reservoir will be shortened and operation and maintenance costs will be unexpectedly increased. Therefore it can be understood that localized dredging is an effective tool to prolong the dam reservoir life and its determined utility and that is why it is currently being used worldwide.

3.3 LOST STORAGE REPLACEMENT TECHNIQUES AND DECOMMISSIONING

3.3.1 Raising dam height

The raising of a dam height is to increase the reservoir capacity in order to compensate the storage loss due to sedimentation. Especially in arid regions, raising the dam height should be seen as a cost-effective method. However, in the long term period it is not a solution for sediment problem but a remedy to store more water. In addition, this method requires very careful engineering and also it causes some problems (Howard, 2000):

●   Socio-economic and political issues related to resettling of people.
●   Increased water losses due to evaporation and seepage.
●   Dam safety aspects which could lead to high raising costs.
●   Impacts of dam use.

3.3.2   Building a new dam

In order to replace the storage loss of an existing dam, a new dam can be built down-stream or upstream of an existing reservoir or on another river. Generally, this practice has been followed in Turkey to replace the lost storage capacity but it is a temporary solution and it is not an environmentally friendly method.

3.3.3   Decommissioning

Decommissioning is removing all the structures of a dam project and so ending the operation life of the dam. Decommissioning of a dam is not a reservoir sediment management technique; on the contrary it is an economical option if the dam's useful life is finished. In other words, if an operation cost of the reservoir is more than the benefits gained from the reservoir, decommissioning is economically an option for further actions.

In addition to economical reasons other reasons to consider dam removal are, according to Howard (2000):

- Water quality improvement.
- Flora and fauna improvement.
- Public Safety Hazard Elimination.
- Aesthetic improvement.
- Existence of an alternative which provides the same advantages as the dam after decommissioning.
- Recreational development.

Chapter 4

Performance of reservoir conservation model

In the previous chapter, methods to achieve "Sustainable Development of Basin Water Storage Capacity" are discussed separately. However, there is an open source package program, RESCON, to examine and compare some sediment evacuation methods and decommissioning, economically and hydraulically. These sediment evacuation methods are; flushing, hydrosuction sediment removal (HSRS), dredging and trucking.

RESCON, which is a World Bank sponsored project, was developed in order to make preliminary decisions for policy makers. RESCON's key algorithm is based on economic optimization and supported by technical evaluation of basic parameters. The economic optimization results determine which sediment management technique is the most viable. In addition, sustainability of the evacuation methods is identified by the program and technical evaluation results can be found. If the sustainable usage is failed then the program computes the annuities for the retirement fund.

4.1 THE GENERAL WORKING PRINCIPLE OF RESCON

RESCON philosophy is actually "Life Cycle Management Approach", which is an alternative of the "Design Life Approach". The Life Cycle Approach's basis is the sustainable usage of projects infinitely opposite to the Design Life Approach, which assumes finite project life. If sustainable usage is not achievable the decommissioning of dam within a finite time is proposed and the necessary annuity for the retirement fund is calculated. The program structure is sketched in Figure 4.1 (Palmieri et al., 2003).

RESCON is a spreadsheet-based program written in Visual Basic programming language and works with macros. There are two sheets to input the required data. Seven classes of data should be input into the first sheet: Reservoir Geometry, Water Characteristics, Sediment Characteristics, Removal Parameters, Economic Parameters, Flushing Benefit Parameters and Capital Investment. In the second sheet, for the selection of a desirable sediment management strategy information about Environmental and Social Safeguard Policies are asked. These input parameters are given in Table 4.1.

Second Sheet: User Input (Environmental and Social Safeguards Page).

If the user is interested in the Environmental and Social Safeguards Policies, this second sheet should be filled in according to the related project data. Otherwise, there are already existing default values and the user should not change these. Default values

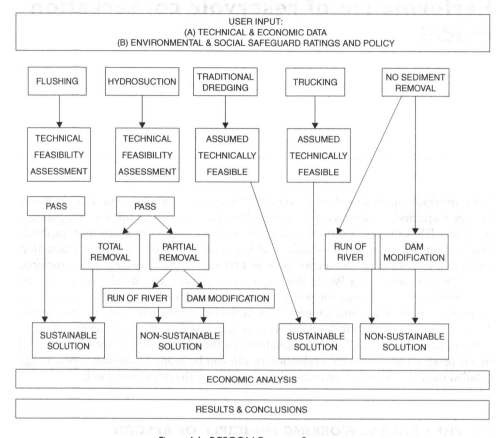

Figure 4.1   RESCON Program Structure

and what these mean are explained in Tables 4.2–4.5, which are taken from an original RESCON program User Input sheet.

As a default value, RESCON takes safeguard rating as "1" and Safety Policy Criteria as "D".

After the above mentioned data is provided, and the results are calculated, the output of the program gives information in different sheets successively:

– Flushing Technical Results
– HSRS Technical Results
– Flushing Technical Calculations
– HSRS Technical Calculations
– Economic Results and Calculations
– Safeguard Results

The above mentioned output values provided as a consequence of a technical and economical optimization. These optimizations' working principle is discussed below.

Table 4.1 First sheet: User Input (Checklist)

Parameter	Units	Description
Reservoir Geometry		
S_o	(m^3)	Original (pre-impoundment) capacity of the reservoir.
S_e	(m^3)	Existing storage capacity of the reservoir.
W_{bot}	(m)	Representative bottom width for the reservoir – use the widest section of the reservoir bottom near the dam to produce worst case for criteria.
SS_{res}		Representative side slope for the reservoir. 1 Vertical to SS_{res} Horizontal.
EL_{max}	(m)	Elevation of top water level in reservoir – use normal pool elevation.
EL_{min}	(m)	Minimum bed elevation – this should be the riverbed elevation at the dam.
EL_f	(m)	Water elevation at dam during flushing – this is a function of gate capacity and reservoir inflow sequence. Lower elevation will result in a more successful flushing operation.
L	(m)	Reservoir length at the normal pool elevation.
h	(m)	Available head – reservoir normal elevation minus river bed downstream of dam.
Water Characteristics		
V_{in}	(m^3)	Mean annual reservoir inflow (mean annual run-off).
Cv	(m^3)	Coefficient of Variation of Annual Run-off volume. Determine this from statistical analysis of the annual run-off volumes.
T	(°C)	Representative reservoir water temperature.
Sediment Characteristics		
ρ_d	(tonnes/m^3)	Density of in-situ reservoir sediment. Typical values range between 0.9–1.35.
M_{in}	(metric tonnes)	Mean annual sediment inflow mass.
Ψ	1600, 650, 300, 180	Select from: 1600 for fine loess sediments; 650 for other sediments with median size finer than 0.1 mm; 300 for sediments with median size larger than 0.1 mm; 180 for flushing with $Q_f < 50$ m^3/s with any grain size.
Brune Curve No	1 2 3	Is the sediment in the reservoir: (1) Highly flocculated and coarse sediment (2) Average size and consistency (3) Colloidal, dispersed, fine-grained sediment.
Ans	3 or 1	This parameter gives the model a guideline of how difficult it will be to remove sediments. Enter "3" if reservoir sediments are significantly larger than median grain size (d_{50}) = 0.1 mm or if the reservoir has been impounded for more than 10 years without sediment removal. Enter "1" if otherwise.
ype	1 or 2	Enter the number corresponding to the sediment type category to be removed by hydrosuction dredging: 1 for medium sand and smaller; 2 for gravel.
Removal Parameters		
HP	1 or 2	Is this a hydroelectric power reservoir? Enter 1 for yes; 2 for no.
Q_f	(m^3/s)	Representative flushing discharge. This should be calculated with reference to the actual inflows and the flushing gate capacities.
T_f	(days)	Duration of flushing after complete drawdown.
N	(years)	Frequency of flushing events (whole number of years between flushing events).
D	(feet)	Assume a trial pipe diameter for hydrosuction. Should be between 1–4 feet.

(Continued)

Table 4.1 Continued

Parameter	Units	Description
NP	1, 2, or 3	Enter the number of pipes you want to try for hydrosuction sediment removal. Try 1 first; if hydrosuction cannot remove enough sediment, try 2 or 3.
YA	Between 0 and 1	Maximum fraction of total yield that is allowed to be used in HSRS operations. This fraction of yield will be released downstream of the dam in the river channel. It is often possible to replace required maintenance flows with this water release. Enter a decimal fraction from 0–1.
CLF	(%)	Maximum percent of capacity loss that is allowable at any time in reservoir for Flushing. For an existing reservoir, this number must be greater than the percentage of capacity lost already. Sustainable solutions will attempt to remove sediment before this percent of the reservoir is filled completely.
CLH	(%)	Maximum percent of capacity loss that is allowable at any time in the reservoir for Hydrosuction. For an existing reservoir, this number must be greater than the percentage of capacity lost already. Sustainable solutions will attempt to remove sediment before this percent of the reservoir is filled completely.
CLD	(%)	Maximum percent of capacity loss that is allowable at any time in the reservoir for Dredging. For an existing reservoir, this number must be greater than the percentage of capacity lost already. Sustainable solutions will attempt to remove sediment before this percent of the reservoir is filled completely.
CLT	(%)	Maximum percent of capacity loss that is allowable at any time in the reservoir for Trucking. For an existing reservoir, this number must be greater than the percentage of capacity lost already. Sustainable solutions will attempt to remove sediment before this percent of the reservoir is filled completely.
ASD	(%)	Maximum percent of accumulated sediment removed per dredging event. Sustainable removal dredging will be subject to this technical constraint.
AST	(%)	Maximum percent of accumulated sediment removed per trucking event. Sustainable removal trucking will be subject to this technical constraint.
MD	(m^3)	Maximum amount of sediment removed per dredging event. The user is warned if this constraint is not met, but the program still calculates the NPV. Use default value unless better information is available.
MT	(m^3)	Maximum amount of sediment removed per trucking event. The user is warned if this constraint is not met, but the program still calculates the NPV. Use default value unless better information is available.
Cw	(%)	Concentration by weight of sediment removed to water removed by traditional dredging. Maximum of 30%. Do not exceed this default unless there is information about the considered reservoir.
Economic Parameters		
E	0 or 1	If the dam being considered is an existing dam enter 0. If the dam is a new construction project enter 1.
c	$(\$/m^3)$	Unit Cost of Construction. The default value given here is a crude estimate based on original reservoir storage capacity. The user is encouraged to replace this value with a project specific estimate.
C2	($)	Total Cost of Dam Construction. This cost is calculated as unit cost of construction times initial reservoir storage volume ($C2 = S_o * c * E$). If you entered $E = 0$ above, your total construction cost will be taken

(Continued)

Table 4.1 Continued

Parameter	Units	Description
		as 0; if you entered E = 1, this cost will be calculated in the above manner.
r	decimal	Discount rate.
Mr	decimal	Market interest rate that is used to calculate annual retirement fund. This could be different from discount rate "r".
PI	($/m^3)	Unit Benefit of Reservoir Yield. Where possible use specific data for the project. If no data is available refer to Volume I report for guidance.
V	($)	Salvage Value. This value is the cost of decommissioning minus any benefits due to dam removal. If the benefits of dam removal exceed the cost of decommissioning, enter a negative number.
omc		Operation and Maintenance Coefficient. This coefficient is defined as the ratio of annual O&M cost to initial construction cost. Total annual O&M cost is calculated by the program as CI = omc * c * So.
PH	($/m^3)	Unit value of water released downstream of the dam in river by hydrosuction operations. This could be zero, but may have a value if the downstream released water is used for providing some of required yield.
PD	($/m^3)	Unit value of water used in dredging operations. This could be zero, but may have a value if settled dredging slurry water is used for providing some of required yield.
CD	($/m^3)	Unit Cost of Dredging – The user is encouraged to input her/his own estimate. Should this be difficult at the pre-feasibility level, enter "N/A" to instruct the program to calculate a default value of the unit cost of dredging. The calculated value is reported in Econ. Results and Conclusion Page.
CT	($/m^3)	Unit Cost of Trucking – The user is encouraged to input her/his own estimate. Should this be difficult at the pre-feasibility level, the default value is recommended.
Flushing Benefits Parameters		
s1	decimal	The fraction of Run-off river benefits available in the year flushing occurs (s1 ranges from 0 to 1).
s2	decimal	The fraction of storage benefits available in the year flushing occurs (s2 ranges from 0 to 1).
Capital Investment		
FI	$	Cost of capital investment required for implementing flushing measures. The cost entered will be incurred when flushing is first practiced.
HI	$	Cost of capital investment to install Hydrosuction Sediment-Removal Systems (HSRS).
DU	Years	The expected life of HSRS.

Table 4.2 Safeguard Ratings for Sediment Management Strategies

Safeguard Ratings for Each Sediment Management Strategy	Safeguard Ratings
No impact and potential benefits	1
Minor impact	2
Moderate impact	3
Significant impact	4

Table 4.3 Safeguard Ratings

Safeguard Policy Criteria	Interpretation	Policy Level
6	No impact and potential benefits	A
7 to 11, with no 3's	Minor impact	B
12 to 15 or at least one 3	Moderate impact	C
16 or higher, or at least 4	Significant impact	D

Table 4.4 Estimated Environmental and Social Impact Levels

Estimated Environmental and Social Impact Levels (Enter 1 to 4)

Possible Strategies	Technique	Natural Habitats	Human Uses	Resettlement	Cultural Assets	Indigenous Peoples	Transboundary Impacts	Total
Nonsustainable (Decommission) with No Removal	N/A	1	1	1	1	1	1	6
Nonsustainable (Decommission) with Partial Removal	HSRS	1	1	1	1	1	1	6
Nonsustainable (Run-off River) with No Removal	N/A	1	1	1	1	1	1	6
Nonsustainable (Run-off River) with Partial Removal	HSRS	1	1	1	1	1	1	6
Sustainable	Flushing	1	1	1	1	1	1	6
Sustainable	HSRS	1	1	1	1	1	1	6
Sustainable	Dredging	1	1	1	1	1	1	6
Sustainable	Trucking	1	1	1	1	1	1	6

Table 4.5 Safeguard Policy Level

Safety Policy Criteria	Policy Level
Maximum allowable environmental and social damage	(A to D)

4.2 WORKING PRINCIPLE OF RESCON FOR TECHNICAL OPTIMIZATION

There are varying sediment management methods as defined in previous sections. However, RESCON considered and analyzed some but not all of the techniques.

These are:

- Flushing
- HSRS
- Traditional Dredging
- Trucking

In addition to these, for the sake of comparison, the "No sediment removal" case is also technically studied by RESCON. The program takes two possibilities for "No Sediment Removal" case after the end of the useful life of dam:

- Run-off river
- Decommissioning

In order to bring the run-off river operations, the reservoir is assumed to be fully depleted with sediment and the dam is functioning to generate power. After these conditions are gained then the existing and fully depleted dam should be maintained for the run-off river operations.

If decommissioning of the dam is considered then the program calculates the most appropriate time, called optimal time, to remove the dam. The optimal time of decommissioning depends on the annual net benefits and salvage value of the dam, which is defined by the user. The cost of dam removal will be calculated by the program with the parameters of the optimal time and the salvage value. Then, in order to accumulate the necessary amount of money the annual retirement fund is calculated by the program.

The formulas related to these will be explained in the Economical Optimization part.

4.2.1 Technical principle of flushing in RESCON

The optimization framework of RESCON in flushing is based on the Atkinson Model (Atkinson, 1996).

Flushing parameters and calculation procedure of the Atkinson Model

According to Atkinson definition, the criteria for feasible reservoir flushing are as follows:

- If the long term balance between sediment flushed and sediment deposited in the reservoir is provided, then the transported sediment through low level outlets by flushing will be sufficient enough for sustainability of the reservoir.
- In order to get a specified volume of the storage, after sediment balance, the remaining volume of sediment will be as small as possible.
- The economical side of the problem should be considered so that the cost of flushing does not exceed the benefits.

In order to understand the above feasibility criteria, the definitions and calculation procedure of Atkinson are reviewed below.

1. **Sediment Balance Ratio (SBR):**

Sediment balance ratio is the ratio of sediment mass flushed annually M_f to the annual sediment mass deposited M_{dep}.

$$SBR = \frac{M_f}{M_{dep}} \tag{4.1}$$

SBR calculation can be done by using the following steps:

i. Derivation of a representative reservoir width from the dam at the flushing water surface elevation according to the reservoir bathymetry:

$$W_{res} = W_{bot} + 2SS_{res}(El_f - El_{min}) \tag{4.2}$$

where, W_{bot} is the width of the reservoir at the bottom and El_f is calculated from outlet sill elevation plus the water depth above that sill at the flushing discharge.

ii. Actual flushing width calculation using a best fit equation resulting from empirical data:

$$W_f = 12.8 \times Q_f^{0.5} \tag{4.3}$$

where Q_f is the flushing discharge (m³/s).

Flushing width is computed by using the above formula which is derived by IRTCES (1985), Jaggi and Kashyap (1984) and Jarecki and Murphy (1963) (Atkinson, 1996).

iii. Take the minimum of W_{res} and W_f as representative width of flow for flushing, W, because the bottom width before impoundment is a limitation for the channel width achieved by flushing.

iv. Estimation of Longitudinal Slope during flushing is:

$$S = \frac{El_{max} - El_f}{L} \tag{4.4}$$

where L is reservoir length and El_{max} is the top water level elevation.

v. Estimation of the parameter Ψ, determined by sediment type, for Q_s prediction in order to use empirical equation developed by the Tsinghua University. This method is obtained on the basis of observations at China reservoirs, which use flushing.

$\Psi = 1600$ for fine loess sediments
$\Psi = 650$ for $D_{50} < 0.1$ mm (median size of sediments are finer than 0.1)
$\Psi = 300$ for $D_{50} \geq 0.1$ mm (median size of sediments are larger than 0.1)
$\Psi = 180$ for low discharge (say less than 50 m³/s) with any grain size

vi. Calculation of the sediment load during flushing.

$$Q_s = \psi \times \frac{Q_f^{1.6} \times S^{1.2}}{W^{0.6}} \text{ (Tones/sec)} \tag{4.5}$$

In order to use this equation S is limited by $0.000006 < S < 0.016$.

In addition, if the reservoir in question is not similar to the Chinese reservoirs studied, Q_s should be reduced by a factor of 3.

vii. Determination of the sediment flushed annually.

$$M_f = 86.400 \times T_f \times Q_s \text{ (tones)} \tag{4.6}$$

where T_f is duration of flushing in days and 86.400 is the number of seconds in a day.

viii. Choose the value of Trap Efficiency (TE) according to Brune's Curve.

Trapping efficiency is the percentage of the trapped sediment related to inflowing sediment. Brune (1953) developed a curve which shows the correlation between reservoir capacity and water inflow with trap efficiency. Actually the Brune Curve consists of three curves which are classified as:

– Highly flocculated and coarse sediment curve.
– Median curve for normal pounded reservoirs and average sediment size.
– Fine sediment.

If sluicing is applied to the reservoir TE is 100%, otherwise by using the Capacity (C) and inflow (I) from the Brune's Curve TE is founded.

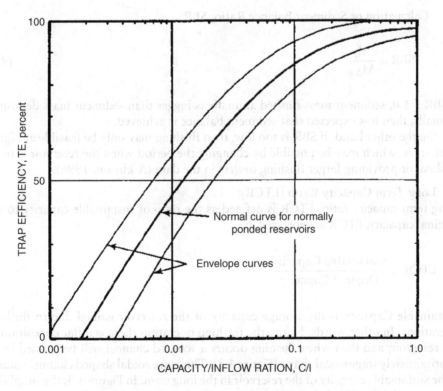

Figure 4.2 Brune's Curve (Brune, 1953, see Atkinson, 1996)

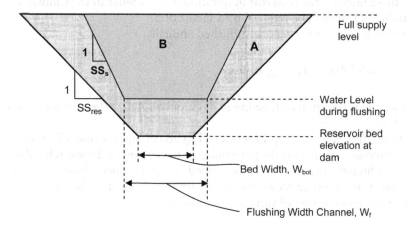

Figure 4.3 Enlarged Section immediately upstream of a dam

ix.   Calculation of the annual sediment mass deposited, M_{dep}

$$M_{dep} = M_{in} \times TE/100 \tag{4.7}$$

x.   Calculation of Sediment Balance Ratio, SBR

$$SBR = \frac{M_f}{M_{dep}} \tag{4.8}$$

If SBR > 1.0, sediment mass flushed annually is bigger than sediment mass deposited annually, then it is expected that sediment balance is achieved.

On the other hand, if SBR is too low, then flushing may only be feasible at higher discharges, which may be possible by changing the period when the reservoir is to be flushed, or providing larger flushing outlets in the dam (Atkinson, 1996).

2.   Long Term Capacity Ratio (LTCR):
Long term capacity ratio, LTCR is defined as the ratio of sustainable capacity to the original capacity. LTCR is expressed as:

$$LTCR = \frac{\text{Sustainable Capacity}}{\text{Original Capacity}} \tag{4.9}$$

Sustainable Capacity is the storage capacity of the reservoir gained due to flushing operations. In other words, before the flushing operation there is a flat deposition in the reservoir, and then when flushing occurs a scoured channel will be formed in an approximately trapezoidal shape (Figure 4.3). This trapezoidal shaped channel volume is the sustainable capacity of the reservoir in the long term. In Figure 4.3, the simplified geometry of the scoured channel can be seen.

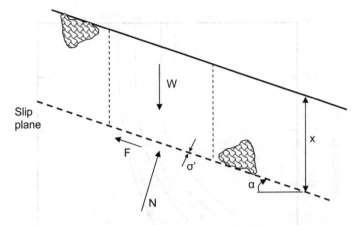

Figure 4.4   Force Balance on a side slope (Atkinson, 1996)

According to the Figure 4.3 the Long Term Capacity Ratio, LTCR, is approximated to:

LTCR = Area B/(Area A + Area B)

LTCR should be calculated by following the steps explained below:

i.   Determination of the scoured valley width at the top water level. Scoured Valley width is actually depending on the W.

$$W_{tf} = W + 2 \times SS_s \times (El_{max} - El_f) \tag{4.10}$$

where SS_s is the representative side slope for the deposits exposed during flushing.

The prediction of side slope studies is essentially based on the force balance shown in Figure 4.4, that simply assuming friction forces parallel to the slope is equal to the down slope gravity forces.

From Figure 4.4, the side slope can be calculated from force equilibrium such that:

In figure W is the weight and expressed as:

$$W = \rho_{bulk} \times g \, (N/m^2) \tag{4.11}$$

in which ρ_{bulk} is bulk density and g is gravitational acceleration.

And N is the normal force which is defined as:

$$N = W \cos \alpha \tag{4.12}$$

in which α is the angle of slope

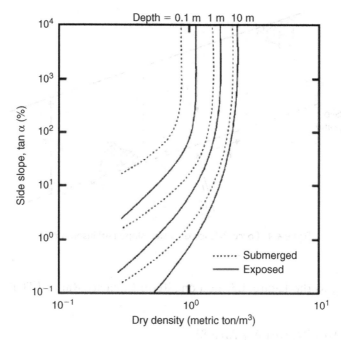

Figure 4.5 Side Slope predictions at the limit of stability (Atkinson, 1996)

And friction force in case no water pressures:

$$F = N \tan \Phi \tag{4.13}$$

where Φ is angle of friction.

$$F = \sigma' \tan \Phi \tag{4.14}$$

where σ' is effective stress.

Then, there are two methods for the prediction of the side slope: the prediction chart and Migniot's equation, which is adopted with multiplier 5 (five) in order to account the difference between submerged and exposed deposits (Atkinson, 1996). The chart and Migniot's equation are below:
Migniot's equation:

$$\tan \alpha = \frac{31.5}{5} \times \rho_d^{4.7} \tag{4.15}$$

where ρ_d is dry density in t/m^3.

ii. Determination of the reservoir width at elevation (El_f) for the assumed simplified geometry.

$$W_t = W_{bot} + 2 \times SS_{res} \times (El_{max} - El_{min}) \tag{4.16}$$

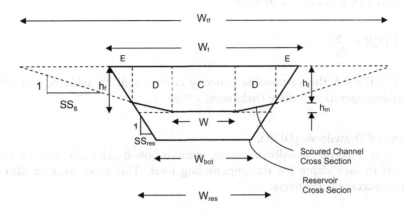

Figure 4.6 Cross section immediately upstream of a dam for simplified reservoir geometry and the scoured channel constricted by reservoir sides (Atkinson, 1996)

iii. If $W_{tf} \leq W_t$ this means that reservoir geometry does not constrict the scoured valley width. Then the cross sectional area of the scoured valley can be calculated by following.

$$A_f = \frac{W_{tf} + W}{2} \times (El_{max} - El_f) \tag{4.17}$$

iv. If $W_{tf} > W_t$ this means that reservoir geometry constrict the scoured valley width. Then cross sectional area of scoured valley will be as shown below.
    According to Figure 4.6:

$$h_m = \frac{W_{res} - W}{2 \times (SS_s - SS_{res})} \tag{4.18}$$

$$h_l = El_{max} - El_f - h_m \tag{4.19}$$

$$h_f = El_{max} - El_f \tag{4.20}$$

Then A_f if the sum of the areas C, D and E.

$$A_f = W \times h_f + (h_f + h_l) \times h_m \times SS_s + h_l^2 \times SS_{res} \tag{4.21}$$

v. Calculation of reservoir cross sectional area is done by using the following formula:

$$A_r = \frac{W_t + W_{bot}}{2} \times (El_{max} - El_{min}) \tag{4.22}$$

vi. Finally LTCR should be defined.

$$LTCR = \frac{A_f}{A_r} \tag{4.23}$$

If $LTCR > 0.5$ the sustainable capacity criteria will be achieved, an effective flushing operation is done (Atkinson, 1996).

3. Extent of Drawdown (DDR):

The extent of reservoir drawdown is unity minus a flow depth ratio, which is flushing water level to flow depth for the impounding level. This ratio gives an idea if the drawdown executed effectively.

$$DDR = 1 - \frac{El_f - El_{min}}{El_{max} - El_{min}} \tag{4.24}$$

The drawdown is insufficient for effective flushing if $DDR < 0.7$ (Atkinson, 1996).

4. Drawdown Sediment Balance Ratio (SBR$_d$):

SBR_d is calculated in a similar way of Sediment Balance Ratio (SBR), but this new ratio is calculated for full drawdown conditions. In other words, in steps (i) and (iv) of SBR calculation use El_f instead of El_{min}. $SBR_d > 1.0$ is preferred (Palmieri et al., 2003).

5. Flushing Channel Width Ratio (FWR):

The flushing width ratio is checked whether the predicted flushing width is greater than the representative bottom width or not.

$$FWR = \frac{W_f}{W_{bot}} \tag{4.25}$$

It will be an important constraint if the FWR is less than one (Atkinson, 1996).

6. Top Width Ratio (TWR):

The ratio of the scoured valley width at the top of the water level to the reservoir actual top width gives the Top Width Ratio, which is used to quantify the side slope. Due to the fact that the side slope can be a constraint for flushing when the FWR value is less than unity or when the scoured valley width is relatively smaller than the actual top width.

$$TWR = \frac{W_{td}}{W_t} \tag{4.26}$$

where W_{td} calculated as follows:

$$W_{td} = W_{bf} + 2 \times SS_s \times (El_{max} - El_{min}) \tag{4.27}$$

If the FWR value is important then the TWR value should exceed 2 in order to overcome the FWR constraint. Otherwise $TWR \approx 1$ is enough (Atkinson, 1996).

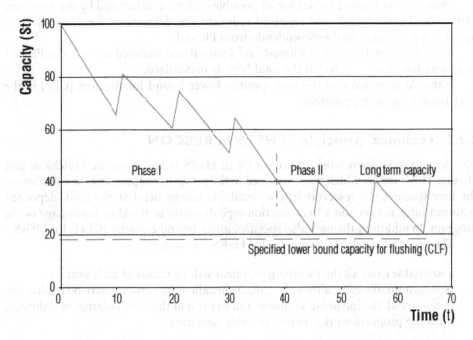

Figure 4.7 Possible Time Path of Remaining Capacity for Flushing

On the other hand, the RESCON model determines the technical feasibility of flushing based on SBR alone (Kawashima et al., 2003). The program takes neither FWR nor TWR into consideration.

The program assumes two phases for flushing operations, namely: Phase I and Phase II. In Phase I, flushing is done periodically until the reservoir reaches its long term capacity (LTC). Then, in order to provide long term capacity in its original level, flushing is done periodically for all subsequently accumulated sediment. In the Figure 4.7, this process is depicted.

The amount of sediment removed by flushing in Phase-II is:

$$LTCR * (So - St) \tag{4.28}$$

where "So − St" is the difference of original storage capacity and remaining storage capacity, in other words accumulated sediment. It is obviously noted that if the remaining storage capacity decreases, the quantity of removed sediment will increase for each cycle.

In addition, if the frequency level of flushing is shorter, then the remaining capacity will be higher than the long term capacity. It could also be discerned from Figure 4.7. However, RESCON does not consider this effect and the optimal cycle length is rather calculated by taking into consideration the determined long term capacity defined at the initial steps of the program.

Then the Net Present Values for all possible cycles are calculated by the program, and optimal cycle length and removed sediment are determined for Phase-II. This optimal cycle is calculated independently from Phase-I.

After that, net benefits of Phase-I and Phase-II are summed up and the Phase-I cycle length is chosen such that the total NPV is maximized.

It should be noted that the user specifies lower bound for flushing (CLF) in the User Input page of the program.

4.2.2 Technical principle of HSRS in RESCON

The optimization framework of RESCON in HSRS is based on the Hotchkiss and Huang (1995) model which was explained in the previous chapter. For the calculation, the user specifies the reservoir length, available energy head at the dam, deposited sediment information and a hydrosuction pipe diameter in the User Input page of the program. In addition, the user also specifies lower bound capacity (CLH) for HSRS.

The program works on two cases for HSRS. These are:

– Sustainable case: all the incoming sediment will be removed each year.
– Non-sustainable case: although HSRS is installed, its capacity may be inadequate to remove all the incoming sediment and to prevent the accumulation of sediment. Then the program works on two possible scenarios.

  • Decommissioning
  • Run-off river

In the sustainable case, according to the specified CLH the program determine the long-term capacity. For the non-sustainable case, decommissioning and run-off river operation scenarios are discussed. As a result, the program reports the optimal timing of HSRS installation time, the amount of sediment removed every year, terminal time for the case of partial removal and also retirement fund for decommissioning case (Kawashima et.al., 2003). Figure 4.8 shows the possible time path.

4.2.3 Traditional dredging and trucking technical principle in RESCON

There are many types of dredging methods and different types of dredging equipment used throughout the world. However, the program uses the aforementioned traditional dredging, removing the silted sediment from reservoir bed by pumping water (Turner, 1996).

The technical feasibility of dredging and trucking does not depend on the sediment removal rate, unlike the previously mentioned methods. They are assumed to always be feasible. Therefore, the user should pay attention to the results and be cautions. Furthermore, in some cases, even though the program claims that the method is feasible; the use of method may not be practical.

The user specifies the following:

– CLD: Maximum percent of the capacity loss that is allowable at any time in the reservoir for Dredging.

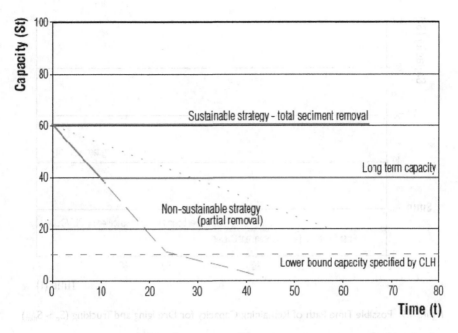

Figure 4.8 Possible Time Path of Remaining Capacity for Hydrosuction

- CLT: Maximum percent of the capacity loss that is allowable at any time in the reservoir for Trucking.
- ASD: Maximum percent of the accumulated sediment removed per dredging event.
- AST: Maximum percent of accumulated sediment removed per trucking event.
- MD: Maximum amount of the sediment removed per dredging event.
- MT: Maximum amount of the sediment removed per trucking event.
- CD: Unit cost of dredging (there is an option in the program that default value can be used by entering "N/A").
- CT: Unit cost of trucking.

During the subsequent execution phase of the program as if the reservoir is new, in Phase-I sediment removal is not considered, as for Phase-II the sediment removal is considered to be constant for each cycle, so that sustainability is provided. Then after economic optimization, S_{min} (lower bound of reservoir capacity) and LTC (long term capacity) are determined according to the optimal duration of Phase-I and optimal cycle length of Phase-II respectively, by the program. It should be noted that Phase-I duration is independent from the Phase-II cycle duration (Figure 4.9).

The optimization can also be done for an existing reservoir. If the existing reservoir capacity, S_e, is lower than optimally determined minimum reservoir capacity, S_{min}, then the optimal time path will be recalculated (Kawashima et al., 2003). An immediate sediment removal occurs and the amount of this initial removal will determine LTC as shown in Figure 4.10.

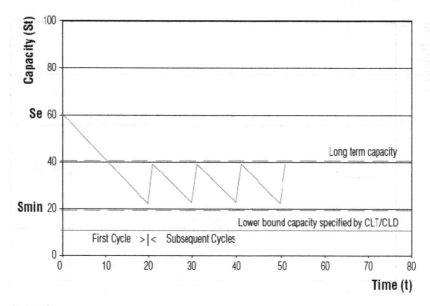

Figure 4.9 Possible Time Path of Remaining Capacity for Dredging and Trucking ($S_e > S_{min}$)

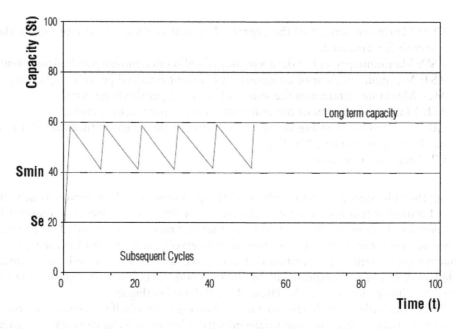

Figure 4.10 Possible Time Path of Remaining Capacity for Dredging and Trucking ($S_e < S_{min}$)

4.3   ECONOMIC OPTIMIZATION WORKING PRINCIPLE OF RESCON

RESCON is a preliminary tool for decision makers to decide whether the investment is feasible or not. RESCON enables them to see the whole picture in advance. From an economical point of view, the RESCON principle is based on maximizing the life-time aggregate Net Present Value. The following optimization is used for all four sediment evacuation techniques mentioned above and also for the "do nothing alternative".

$$\text{Maximize} \sum_{t=0}^{T} NB_t \cdot d^t - C2 + V \cdot d^T \tag{4.29}$$

$$\text{Subject to: } S_{t+1} = S_t - M + X_t \tag{4.30}$$

where:
- NB_t = annual net benefits in year t
- d = discount factor (defined as $1/(1+r)$, where r is rate of discount)
- C2 = initial cost of construction for proposed dam (=0 for existing dam)
- V = salvage value
- T = terminal year
- S_t = remaining reservoir capacity in year t
- M = trapped annual incoming sediment
- X_t = sediment removed in year t

(Kawashima et al., 2003)

According to the above optimization, the most beneficial sediment evacuation method will be chosen. The highest value of the objective function gives the best solution. The Net Benefit calculation is different for different sediment evacuation methods. These are explained below.

The parameters used in the calculation of Annual Net Benefit equations are defined below:

- P1 = Unit Value of Reservoir Yield
- W_t = Annual Reservoir Yield, which is a function of S_t
- S_t = Remaining Reservoir Capacity
- X_t = The Amount of Sediment Removed
- Y_t = The needed water for sediment removal
- C1 = Annual operations and maintenance cost
- PH = Downstream value of water used during hydrosuction
- PD = Downstream value of water used during dredging
- CH = Unit Cost of sediment evacuating with HSRS
- CD = Unit Cost of sediment evacuating with dredging
- CT = Unit Cost of sediment removal with trucking
- FI = Capital Cost of installing a Flushing system

There are three different Net Benefit calculations for the Flushing method. If the amount of evacuated sediment is zero, then the Net Benefit is equal to the net income minus net outcome. In other words, the unit value of reservoir yield multiplied by the annual reservoir yield, minus annual operations and maintenance cost. Moreover, if the amount of evacuated sediment is bigger than zero, then in first flushing the capital cost of installing a flushing system will be an outcome for Net Benefit.

For Flushing:

$$
NB_t = \begin{cases}
P1 \cdot W(S_t) - C1 & \text{if } X_t = 0 \\
P1 \cdot [s1 \cdot W(0) + s2 \cdot (W(S_{t+1}) - W(0))] & \text{if } X_t > 0, \\
\quad - C1 - F1 & \text{First Flushing} \\
P1 \cdot [s1 \cdot W(0) + s2 \cdot (W(S_{t+1}) - W(0))] & \text{if } X_t < 0, \\
\quad - C1 & \text{Subsequent} \cdot \text{Flushing}
\end{cases}
$$

(4.31)

For HSRS:

$$
NB_t = P1 \cdot W(S_t) - (P1 - PH) \cdot Y_t - C1 - CH \cdot X_t \tag{4.32}
$$

Cost of sediment evacuating with HSRS and water used for HSRS will be taken as outcome. On the other hand, if water used during HSRS has a downstream value, it will be income.

For Traditional Dredging:

$$
NB_t = P1 \cdot W(S_t) - (P1 - PD) \cdot Y_t - C1 - CD(X) \cdot X_t \tag{4.33}
$$

Its logic is exactly the same as HSRS: Cost of sediment evacuating with dredging and water used for dredging will be taken as outcome. On the other hand, if water used during dredging has a downstream value, it will be income.

For Trucking:

$$
NB_t = P1 \cdot W(S_t) - C1 - CT \cdot X_t \tag{4.34}
$$

(with $W_t = 0$ if $X_t > 0$)

Basically since the reservoir should be empty for the trucking method there is no income. As an outcome cost of sediment removal with trucking will be taken.

For no removal:

$$
NB_t = P1 \cdot W_t - C1 \tag{4.35}
$$

The Net Benefit is obviously seen from the equation: net income minus net outcome.

In addition to these, the Annual Reservoir Yield (W_t) which is a function of the Remaining Reservoir Capacity (S_t) is calculated via Gould's Gamma Function (RESCON Manual Vol-I, 2003).

$$
W_t = \frac{4 \cdot S_t \cdot V_{in} - Zpr^2 \cdot sd^2 + 4 \cdot Gd \cdot sd^2}{4 \cdot \left(S_t + \dfrac{Gd}{V_{in}} \cdot sd^2 \right)} \tag{4.36}
$$

where:

V_{in} = incoming flow volume (annual run-off)

sd = standard deviation of incoming flows (annual run-off)

Z_{pr} = standard normal variate of p%

Gd = adjustment factor to approximate the Gamma distribution (offset from the normal distribution)

Since this economical function is another academic area and its understanding is not important for the usage of RESCON, its details are not discussed in this study. It is included only for information. However, it should be noted that W_t (annual reservoir yield) is calculated for every step, t, in the economic model (Palmieri et al., 2003).

In the case of sediment removal, the required water for successful sediment evacuation for each and every sediment evacuation method used in the economical model of RESCON is discussed below.

4.3.1 Flushing

During Flushing, since the reservoir will be emptied, the water yield, W_t, is determined as follows:

$$W_t = s1 \cdot W(0) + s2 \cdot (W(S_{t+1}) - W(0)) \tag{4.37}$$

in which:

s1 = the fraction of Run-of river benefits available in the year flushing occurs

W(0) = water yield from Run-of river project

s2 = the fraction of the storage benefits available in the year flushing occurs

$W(S_{t+1})$ = water yield from storage capacity after flushing.

4.3.2 HSRS

The required water for the evacuation of deposited sediment from a reservoir by HSRS, Y_t, is expressed as:

$$Y_t = \left(\frac{Q_m}{Q_s}\right) \cdot X_t \tag{4.38}$$

where:

Q_m = mixture flow rate

Q_s = sediment flow rate

X_t = sediment removed in year t.

4.3.3 Dredging (Traditional)

The required volume of water for removing a specified volume of sediment, Y_t, is expressed as:

$$Y_t = \left(\frac{100 * 2.65}{C_w}\right) \cdot X_t \tag{4.39}$$

where C_w is specified by the user, which is the concentration of sediment weight to water removed.

4.3.4 Trucking

The W_t is assumed to be zero for simplicity. Because for trucking process the reservoir should be empty and there is no need for water.

Therefore, RESCON calculates a retirement fund and a terminal time, if the reservoir will be decommissioned. The accumulation of this Retirement Fund in the duration of Terminal Time provides the necessary cost amount for the decommissioning of a dam at the end of the reservoir life. The program calculates the retirement fund by using the equation below:

$$k = -mV/((1+r)^T - 1) \tag{4.40}$$

where:
k = annual retirement fund
m = rate of interest (different from the discount rate r)
V = Salvage Value
T = terminal year

As a result of this Net Benefit and Retirement Fund calculations, program gives the solution in two forms:

1 If the reservoir performs its task forever, this case is called *"Sustainable"*.
2 If the reservoir performs its task within a finite period, this case is called *"Non-sustainable"*.
   There are two possibilities for non-sustainable case.

   a) Decommissioning of a dam at its Terminal Time
   b) After the siltation of a dam the reservoir can be used as a run-off river project.

RESCON economic calculation based on the below presented formulas and relationship.

4.3.5 Unit cost of the evacuation methods used in RESCON

• Hydrosuction (HSRS) Unit Cost is determined as:

$$CH = \frac{HI}{DU \cdot Q_s} \tag{4.41}$$

where:
CH = unit cost of hydrosuction
HI = cost of capital investment to install HSRS
DU = the expected life of HSRS
Q_s = the annual maximum transport rate

The HI and DU is specified by the user in the User Input Page of the RESCON excel sheet. The annual maximum transport rate, Q_s, is automatically calculated by the program and this value can be found in the HSRS Technical Calculations excel page of the program.

- Dredging Unit Cost is determined as follows:

    If $X < 150,000 \, \text{m}^3$ $CD(X) = 15.0$

    If $X > 16,000,000 \, \text{m}^3$ $CD(X) = 2.0$

    Else $CD(X) = (6.61588727859064) * (X/(10^6))^\wedge -0.431483663524377$

$$(4.42)$$

where:

    X = amount of sediment dredged per cycle (m^3)
    CD = unit cost of dredging $(\text{US\$/m}^3)$

Although the program calculates the unit cost of dredging by using the above formula, the program encourages the users to enter their own values. In the User Input page of the RESCON program, the CD value area is ready for the user to input the specific cost data of dredging. From the above formula it is obviously seen that the dredging cost decreases if the volume of removed sediment increase.

- Construction Unit Cost is determined as follows:

    If $So > 500,000,000 \, \text{m}^3$ $c = \text{US\$0.16/m}^3$

    Else $c = 3.5 - 0.53 * LN(So/1000000)$

$$(4.43)$$

where:

    So = original storage capacity
    c = unit construction cost

The default unit cost of construction is calculated by program if the users should not be able to enter their own value.

From the equation it is understood that the unit cost of construction decreases if the original storage volume increases.

- Annual operations and maintenance cost is determined below:

    $$C1 = omc * c * So \qquad\qquad (4.44)$$

where:

    $C1$ = annual operations and maintenance cost, US\$,
    c = unit cost of dam construction, US\$/ m^3
    omc = operations and maintenance coefficient

The operations and maintenance coefficient, omc, is provided by the user in the User Input page of the program.

4.4 EVALUATION AND COMMENTS ABOUT ECONOMIC RESULTS OF RESCON

One of the output pages of RESCON is the "Economic Results and Calculations" page. Although the alternatives have different NPV calculation method, they are compared to each other on the same basis. In order to understand how this comparison works out, the difference between "Design Life" and "Life Cycle" terms should be known very well. RESCON tries to determine how to manage the facility in the most optimal way by using infinite time in other words in perpetuity.

In order to compare the alternatives on the same basis, the RESCON uses 300 years to define perpetual time for all the sustainable projects. Therefore, the sustainable optimal management procedures for removing sediment are defined. The NPV calculation is done for 300 years; since the life cycles of each method are different, the time of removal are different. For example, in 300 years the dredging method may be used 10 times but the trucking method may be used 5 times.

On the other hand, in the calculation of non-sustainable cases, which means the reservoir eventually silts up completely and has to be decommissioned or used as a run-off river, a finite design life is used. In addition, for a decommissioning case, an investment fund is determined to save enough money for decommissioning for the future generation.

Chapter 5

Cost analyses

5.1 CASE STUDIES FROM TURKEY

The case studies using the RESCON program are Cubuk Dam-I, Ivriz Dam, Borcka Dam and Muratlı Dam. The detailed information about these case studies can be found in the following sections. The reason for using only Turkey dams is that we have chance to gained enough data to run RESCON programme.

5.2 HISTORICAL BACKGROUND OF DAM CONSTRUCTION IN TURKEY: EXPERIENCE, LESSONS, MALPRACTICES

In Turkey most of the dams can be considered as large dams and most of these large dams have been constructed for irrigation and domestic water supply. However, since there has recently been increased energy need, as in other countries, hydropower dams are becoming important and their construction is taken into consideration.

In addition, sediment studies related to sustainability are quite rare in Turkey. There is no sediment removal operation made on a large scale in any reservoir. Practiced sediment removal operations are only for clearing around water intake structures or similar local operations. There are some studies done by the State Hydraulic Works but they are generally related to the sediment problem in local areas and written to advise sediment prevention strategies for that region. Turkey is a country having varied wide areas subject to erosion. Green cover in Turkey is not enough to prevent sediment coming into reservoirs. Large seasonal flows also threat watershed and may increase sedimentation.

Moreover, at the design level of a dam in Turkey, the design life approach is used and life cycle management is not considered.

Therefore, solutions should be found for sustainable management of existing reservoirs as well as new ones.

5.2.1 Reservoir sedimentation in Turkey

The siltation in dams is a growing, and dangerous, problem worldwide as much as for Turkey. The life of a dam directly depends on the sedimentation. Besides, the magnitude of sedimentation depends on several natural factors, such as: climate, geographic and geological conditions. As a result, the life of a dam is indirectly affected by natural conditions.

Turkey has a semi-arid climate and this is the most unpredictable condition for sediment production. According to the world records the measurements of sediment production for semi-arid conditions in the world are as high as 6000–8000 m^3/km^2/year. Therefore, the deposited sedimentation in Turkey reservoirs should be highly considered.

5.2.2 River basins

There are 26 catchment areas in Turkey. The basins of Turkey (Figure 5.1) and the catchment areas used in the case studies are given below (Figures 5.2–5.6).

5.2.3 Measurement of deposition in reservoir lakes

State Hydraulic Works (DSI), which works cooperatively with other governmental institutions, is the main resource used to gather sediment related data in Turkey. The method to carry on the business is summarized below.

When a dam construction is planned, the DSI requests sediment data from EIE (Directorate of Electrical Power Resources Survey and Development Agency) if EIE has a gauging station in that region or neighbourhood. If there is no gauging station sediment measurements are taken by the DSI in that region, in a frequency satisfying precision of sediment yield for a period (it may be daily, weekly or monthly), because the precision of sediment yield may be important for small structures like weirs, run-off river power plants, etc. If previously taken sediment measurement data is not available, sediment data of the dams or water structures previously constructed in that region are used with some approximation. Finally, if approximation is not possible, the approximate value for sediment yield is assumed using erosion or sediment yield maps. Dead volume of a reservoir is calculated assuming a 50 year economic life for a dam. Annual sediment yield obtained for that dam is multiplied by 50 in order to obtain the volume of sediment which would deposit in 50 years. This calculated volume is allocated as the dead volume for that dam.

There are 3 sampling methods practiced by DSI. These are:

Point Sampling Method
Point Integration Method
Depth Integration Method

Most of the time the depth integration method is used by the DSI. It obtains vertical variation of suspended sediment concentration at a river section (DSI report, 2005).

For suspended sediment sampling US.P-46 and US.P-46R types of samplers are used for point integration method and US.DH-48, US.D-49 and US.D-43 types of samplers are used for depth integration method.

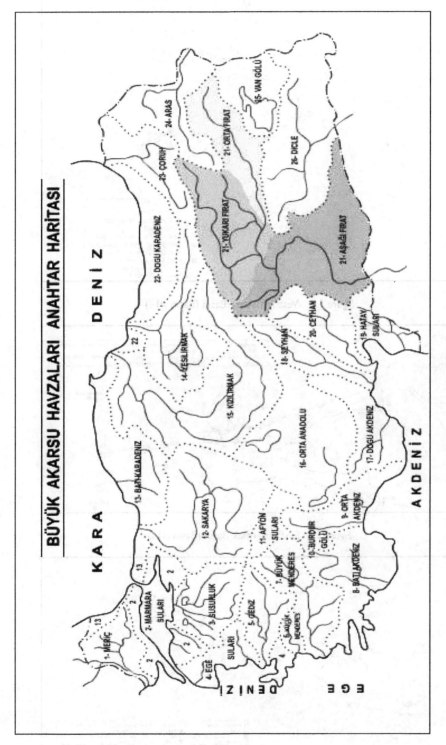

Figure 5.1 Layout of the Basins of Turkey

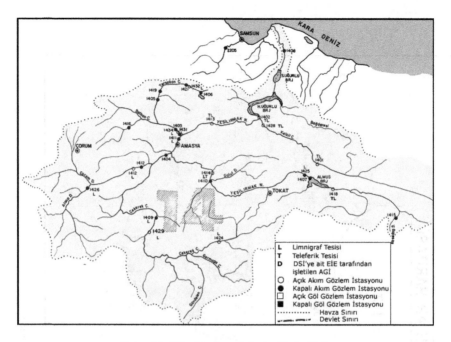

Figure 5.2 West Yesilirmak Basin (Basin #14)

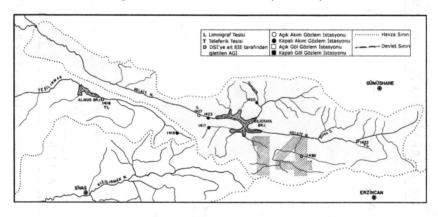

Figure 5.3 East Yesilirmak Basin (Basin #14)

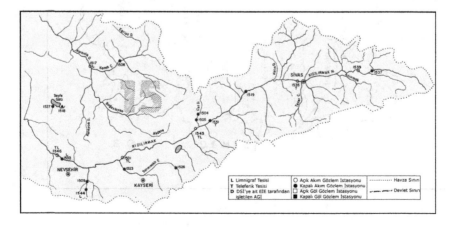

Figure 5.4 South Kizilirmak Basin (Basin #15)

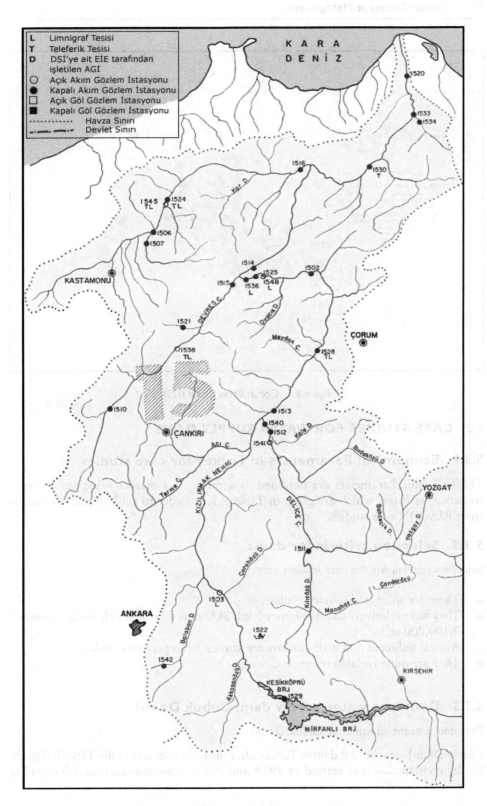

Figure 5.5 North Kizilirmak Basin (Basin #15)

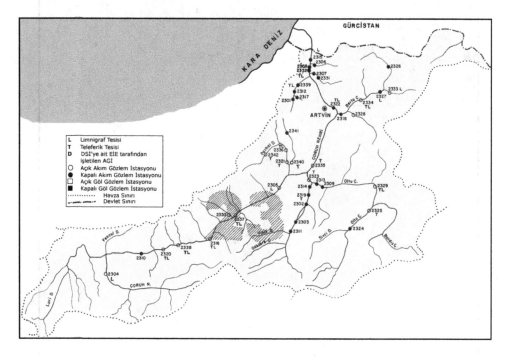

Figure 5.6   Coruh Basin (Basin #23)

5.3   CASE STUDIES FOR WATER SUPPLY DAMS

5.3.1   Economical Parameters in Turkey for case studies

The Economic Parameters are obtained as a result of a market value survey from construction firms, which are given in Table 5.1, in order to achieve realistic results from RESCON case studies.

5.3.2   Selection criteria for dams

Selection criteria for the case studies are:

- There are some preliminary studies.
- They have relatively small volumes (such as Cubuk I Dam which has a volume of 7,100,000 m³).
- Annual sediment and water inflow are known from previous studies.
- They are quite isolated reservoirs.

5.3.3   Domestic water supply dam, Çubuk Dam-I

Past and present situation of Cubuk Dam-I

Çubuk Dam-I was the first dam in Turkey after the establishment of the Turkish Republic. Its construction was started in 1929 and it was commissioned on 3 November

Table 5.1 Economic Parameters of Turkey

Parameter	Unit	Type	Value*
Discount Rate (r)	%	Hydroelectric Power Dam	10
Discount Rate (r)	%	Domestic Water Supply	10
Discount Rate (r)	%	Irrigation	10
Market Interest Rate (Mr)	%	–	10
Unit Benefit of Reservoir Yield (PI)	($/kWh)	Hydroelectric Power Dam	0.069
Unit Benefit of Reservoir Yield (PI)	($/m³)	Domestic Water Supply	0.93
Unit Benefit of Reservoir Yield (PI)	($/m³)	Irrigation	0.39
Salvage Value (V)	($)		Variable according to dam
Operation and Maintenance Coefficient (omc)	%	–	1.0
Unit Value of Water used in HSRS (PH)	($/m³)		0.001
Unit Value of Water used in Dredging (PD)	($/m³)		0.001
Unit Cost of Dredging (CD)	($/m³)		15.00
Unit Cost of Trucking (CT)	($/m³)		4.00

(*) Note: The exchange rate is taken as $1 = 1.4 YTL

Figure 5.7 A view from Çubuk Dam-I

1936 (Figure 5.7). Çubuk Dam-I is located on the Çubuk stream and 12 km north of the Ankara city. It is a concrete gravity dam and its reservoir area is approximately 0.94 km². The project cost was 3,500,000 TL in 1936.

The design of the reservoir and the appurtenant structures was made by Prof. Dr. Walter Kunze with the collaboration of DSI engineers. In addition, he worked as a consultant during the construction stage.

The main characteristics of Çubuk Dam-I are listed below.
Hydrological Data:

- Drainage Area: 720 km^2
- Mean Annual Precipitation: 450 mm
- Estimated annual average run-off: 140×10^6 m^3

Reservoir Data:

- Active volume: 10×10^6 m^3
- Gross reservoir capacity: 12.5×10^6 m^3
- Maximum reservoir volume: 13.5×10^6 m^3
- Normal water surface elevation: 906.61 m
- Maximum water surface elevation: 907.61 m
- Maximum reservoir surface area: 16.5 km^2

Dam:

- Type: Concrete gravity, circular axis
- Height above lowest foundation: 58.00 m
- Height above ground level: 25.00 m
- Crest elevation: 908.61 m
- Crest length: 250 m
- Crest width: 4 m

Spillway Data:

- Type: Gated, 5 span
- Crest length: 2-span part, 16.40 m
               : 3-span part, 19.88 m
- Crest elevation: 905.61 m
- Type of gates: Electrical and manually operated taintor (radial) gates
- Combined maximum discharge capacity: 227 m^3/sec

The purpose of the Çubuk Dam-I was not only to supply domestic and industrial water to the city but also to control floods. However, because of siltation the reservoir cannot maintain its function (Figure 5.8). It has been used only for recreational purposes since the reservoir lost its functionality.

In the literature there are some discrepancies about the amount of deposited sediment in Çubuk Dam-I. In this study, for the capacity loss, Yılmaz (2003)'s study is taken into account corresponding to 50% storage loss in Çubuk Dam-I due to deposited sediment.

Nowadays, in order to gain the lost reservoir capacity, Cubuk Dam-I is emptied and the deposited sediment is evacuated. The photos after evacuation process can be seen in Figures 5.9–5.14.

Figure 5.8 Siltation in Çubuk Dam-I before discharging

Figure 5.9 Siltation level in Çubuk Dam-I after evacuation

Figure 5.10 Reservoir surface area after the discharging of Cubuk Dam-I

Figure 5.11 Side slopes of Çubuk Dam-I after evacuation

Figure 5.12 The deposited sediment type in Cubuk Dam-I

Figure 5.13 Siltation in bottom outlet of Cubuk Dam-I

Figure 5.14 Downstream view of Cubuk Dam-I

Çubuk Dam-I parameters and RESCON analyze

The case studies that were examined were previously performed by Çetinkaya (2006). The necessary technical parameters for RESCON were analyzed in detail by Çetinkaya with an extreme difficulty in gathering them. This study is an effort to improve and enhance his work. Therefore, the technical parameters in Çetinkaya's study are checked from related references and then revised if necessary. Then, they are used to examine the aforementioned cases. The parameters used in RESCON in the user input page are made available in Table 5.2.

Kılıç determined the characteristics of sediment (permeability, grain size, percentage, etc.) in laboratory experiments by taking different samples from the reservoir basin of Çubuk Dam-I. As a result of the analyses, Kılıç reported that since grain size is smaller than 0.147 mm (100 no. sieve diameter) and reservoir size is small, the reservoir can be cleared by mechanical mixing method.

The CaCO$_3$ rate, which is 10–16% in sediment, increases the probability of solidification of deposited material under a particular depth. However, it is impossible to use this sediment as a normal construction material because of the handling difficulty of the deposited sediment which is high plasticity inorganic clay (Kılıç, 1986). In addition, the cleaning of sediment is very difficult because of the mentioned reason. On the other hand, the low specific weight of sediment indicates that cleaning process can be done easily with mechanical mixing. Furthermore, because the type of clay is

Table 5.2 User Input Pages of Cubuk Dam-I

Parameter	Units	Value
Reservoir Geometry		
S_0	(m^3)	7,100,000
S_e	(m^3)	3,550,000
W_{bot}	(m)	57.0
SS_{res}	–	1.0
EL_{max}	(m)	907.6
EL_{min}	(m)	882.6
EL_f	(m)	895
L	(m)	6500
h	(m)	25.0
Water Characteristics		
V_{in}	(m^3)	28,000,000
C_v	(m^3)	0.1
T	(°C)	10.0
Sediment Characteristics		
ρ_d	(tones/m^3)	1.80
M_{in}	(metric tonnes)	81,000
ψ		180
Brune Curve No		3
Ans	–	3
Type		1
Removal Parameters		
HP		2
Q_f	(m^3/s)	27
T_f	(days)	5
N	(years)	1
D	(feet)	4.0
NP		3
YA		0.1
CLF	(%)	60
CLH	(%)	60
CLD	(%)	60
CLT	(%)	60
ASD	(%)	90
AST	(%)	90
MD	(m^3)	1,000,000
MT	(m^3)	3,600,000
C_w	(%)	30
Economic Parameters		
E	–	0
C	($/m^3)	2.46
C2	($)	0
R	(decimal)	0.1
Mr	(decimal)	0.1
PI	($/m^3)	0.93
V	($)	4,500,000
omc	–	0.01
PH		0.001
PD		0.001
CD	($/m^3)	15.00
CT	($/m^3)	4.00
Flushing Benefits Parameters		
s1	(decimal)	0.9
s2	(decimal)	0.9
Capital Investment		
FI	($)	2,000,000
HI	($)	1,000,000
DU	(years)	10

montmorillonit and illit, it cannot be used for the manufacture of tile or brick (Kılıç, 1986). The sediment characteristics coming from the basin is an important parameter for the purposes of cautions to decrease the rate of siltation. By taking into consideration these circumstances, the inflow coming from the basin, and especially from Çubuk Dam-II, should be rested before coming into the reservoir (Kılıç, 1986).

Çubuk Dam-I was built in 1936 and Çubuk Dam-II was built in 1963. Between 1936 and 1963, the sediment yield value of the basin was 372 ton/year/km^2. On the other hand, after the construction of Çubuk Dam-II and between years 1964–1983 the sediment yield value of the basin was 350 ton/year/km^2. As a result of Kılıç analyses, Çubuk Dam-II does not affect the sediment yield value significantly. Therefore in this study the effect of Çubuk Dam-II in sediment yield value is neglected.

The Sediment Characteristics in the Input Sheet of RESCON is deeply studied in this study. According to RESCON, the density value of in-situ reservoir sediment should be between 0.9–1.35. However, in this study for Çubuk Dam-I the density of sediment is taken as 1.8 tonnes/m^3 (Kılıç) because the deposited sediment has not been cleared for years. Therefore the density of sediment in Çubuk Dam-I is higher than default RESCON values.

One of the Sediment Characteristics, Ψ value, is taken as 180 because Q_f value is smaller than 50 m^3/s. The Brune Curve Number in the input sheet is 3, which means the sediment in the reservoir is colloidal, dispersed and fine grained. In addition, "Ans" value is taken as 3, because the reservoir has been impounded for 60 years without sediment removal. For Hydrosuction Dredging, the sediment type is 1 which is for medium and smaller sand.

Although the default values of CLF (%), CLH (%), CLD (%) and CLT (%) in RESCON is 100%, in this study they are taken as 60% – because Çubuk Dam-I is not operated and 50% is already filled with sediment.

The unit benefit of reservoir yield is taken from ASKİ tariff dated 06.03.2008.

Case study of Çubuk Dam-I for flushing feasibility

Analyses, according to Atkinson's calculation procedure, of Çubuk Dam-I is presented in the following sections. All criteria are discussed in order to decide if flushing is appropriate for Çubuk Dam-I. The necessary data for analysis is listed below.

$$C_o = 7.1 \, mm^3$$
$$L = 6,500 \, m$$
$$El_{max} = 907.6 \, m$$
$$El_{min} = 882.6 \, m$$
$$W_{bot} = 57 \, m$$
$$SS_{res} = 1{:}1$$
$$V_{in} = 65.5 \, mm^3$$
$$M_{in} = 81,000 \, t$$
$$Q_f = 27 \, m^3/s$$
$$T_f = 5 \, days \, (120 \, hours)$$
$$El_f = 895 \, m \text{ (assumed due to not knowing sill elevation)}$$
$$\rho_d = 1.35 \, t/m^3$$
$$\tan \alpha = (31.5/5) \times 1.35^{4.7} = 25.82 \text{ (corrected divided by 10)} = 2.582$$
$$SS_s = 1/\tan \alpha = 0.387$$

1. SBR Calculation

i. The representative reservoir width at the flushing water surface elevation:

$$W_{res} = 57 + 2 \times 1 \times (895 - 882.6) = 81.8\,m$$

ii. Actual Flushing width

$$W_f = 12.8 \times 27^{0.5} = 66.5\,m$$

iii. Take the minimum of W_{res} and W_f as the representative width

$$W = 66.5\,m$$

iv. The longitudinal slope during flushing

$$S = \frac{907.6 - 895}{6500} = 0.0019$$

v. The value of Ψ is 180.
vi. The sediment load during flushing is:

$$Q_s = 180 \times \frac{27^{1.6} \times 0.0019^{1.2}}{66.5^{0.6}} = 1.556\,t/s$$

Then this value is reduced by a factor of three according to Atkinson's criteria, since the reservoir in question is not similar to Chinese reservoirs.

$$Q_f = 0.519\,t/s$$

vii. Sediment Mass flushed annually is:

$$M_f = 86,400 \times 5 \times 0.519 = 224,168\,t$$

viii. From Brune's median curve is used and capacity inflow ratio is:

$$C_o/V_{in} = 7.1/65.5 = 0.108 \text{ then } TE = 88\%$$

ix. Sediment mass deposited annually is:

$$M_{dep} = 81,000 \times 0.88 = 71,280\,t$$

x. SBR $= 224,168/71,280 = 3.14$

The SBR value is bigger than unity, this means that the sediment mass flushed annually is bigger than sediment mass deposited annually. It is expected to have successful flushing operation.

2. LTCR Calculation:

i. The Scoured valley width is:

$$W_{tf} = 66.5 + 2 \times 0.387 \times (907.6 - 895) = 76.25 \, m$$

ii. The reservoir width at this elevation is:

$$W_t = 57 + 2 \times 1 \times (907.6 - 882.6) = 107 \, m$$

iii. Since $W_{tf} \leq W_t$ then scoured valley cross sectional area is:

$$A_f = ((76.25 + 66.5)/2) \times (907.6 - 895) = 899 \, m^2$$

iv. Reservoir cross section area is:

$$A_r = ((107 + 57)/2) \times (907.6 - 882.6) = 2,050 \, m^2$$

v. Finally Long Term Capacity Ratio is:

$$LTCR = 899/2,050 = 0.44$$

The LTCR value 0.44 is not bigger than 0.5 thus the operation wouldn't be effective regarding to this criteria.

3. Extent of Drawdown (DDR)
The DDR is:

$$DDR = 1 - \frac{895 - 882.6}{907.6 - 882.6} = 0.504$$

This value is not bigger than 0.7 criteria, suggesting that the flushing is inefficient according to this criterion.

4. Drawdown Sediment Balance Ratio (SBR$_d$)

i. The representative reservoir width at the flushing water surface elevation is:

$$W_{res} = 57 + 2 \times 1 \times (895 - 895) = 57 \, m$$

ii. Actual Flushing width is:

$$W_f = 12.8 \times 27^{0.5} = 66.5 \, m$$

iii. Take the minimum of W_{res} and W_f as the representative width:

$$W = 57 \, m$$

iv. The longitudinal slope during flushing:

$$S = \frac{907.6 - 882.6}{6500} = 0.0038$$

v. The value of Ψ is 180.
vi. The sediment load during flushing is:

$$Q_s = 180 \times \frac{27^{1.6} \times 0.0038^{1.2}}{57^{0.6}} = 3.869\,t/s$$

Then this value is divided into three since the conditions are not same as those in China.

$$Q_f = 1.290\,t/s$$

vii. Sediment Mass flushed annually is:

$$M_f = 86,400 \times 5 \times 1.290 = 557,280\,t$$

viii. From Brune's median curve is used and capacity inflow ratio is $C_o/V_{in} = 7.1/65.5$ $= 0.108$ then $TE = 88\%$
ix. Sediment mass deposited annually is:

$$M_{dep} = 81,000 \times 0.88 = 71,280\,t$$

x. $SBR = 557,280/71,280 = 7.82$

The SBR value is extremely bigger than unity. For full drawdown conditions, it is expected to achieve successful flushing.

5. Flushing Channel Width Ratio (FWR)
The flushing width ratio is:

$$FWR = \frac{66.5}{57} = 1.17$$

This ratio is not a constraint because of being higher than one. Since the predicted flushing width is higher than representative bottom width of the reservoir.

6. Top Width Ratio (TWR)
In order to calculate the top width ratio, first W_{td} should be calculated. W_{bf}, bottom width of scoured valley at full drawdown, should be taken as the minimum of W_f (66.5 m) and W_{bot} (57 m).

$$W_{td} = 57 + 2 \times 0.387 \times (907.6 - 882.6) = 76.35$$

$$TWR = \frac{76.35}{107} = 0.714$$

where W_{td} calculated as follows:
Since FWR is not a constraint, $TWR = 0.714 \approx 1$ is enough.

As a result of Atkinson's calculation procedure flushing of Çubuk Dam-1 will not be effective. Although SBR value is sufficiently large, LTCR value does not meet the criterion for effective flushing operation. If both SBR value and LTCR values meet the Atkinson criteria, the other criteria will be examined.

RESCON User Inputs and Results of Cubuk Dam-I

RESCON optimization results by using Technical and Economical Parameters of Çubuk Dam-I are given below.

The economic results should be evaluated so that the aggregate net benefit of non sustainable solutions is calculated according to their finite design life. However, the aggregate net benefit of sustainable solutions is calculated according to their life cycle principle which has an infinite lifetime but is assumed at 300 years by the program.

As a result of optimization, for Çubuk Dam-I, the highest net aggregate benefit, $231,507,568.02, is achieved by using HSRS technique, which should be seen from Tables 5.3 and 5.4. In this analysis the real water yield value, 0.93 $/m^3, is used for optimization. On the other hand, the value of water is a controversial issue, such that in some areas water is free of charge. Therefore, in this optimization procedure of RESCON the net aggregate benefit differentiates but it is highly dependent on water yield value and it should not be taken as net benefit.

In order to sustain the dam HSRS long term capacity, which is 3,550,000 m^3 (Table 5.8), the approximate number of years result is 1 year (Table 5.9). The long term capacity ratio is depends on the CLD (%) value in the input sheet of RESCON. CLD default value is 100%, however since the Çubuk Dam-I has not been in operation because of sedimentation, and it is already filled by 50%, CLD value is taken as 60%. As a result of this assumption LTCR value decreases for HSRS method.

Table 5.3 Economic Results Summary of Çubuk Dam I

Possible Strategies	Technique	Aggregate Net Present Value
Do nothing	N/A	230,041,523.91
Nonsustainable (Decommissioning) with Partial Removal	HSRS	Partial Removal with HSRS is technically infeasible, see Total Removal with HSRS
Nonsustainable (Run-off River) with No Removal	N/A	230,042,358.29
Nonsustainable (Run-off River) with Partial Removal	HSRS	Partial Removal with HSRS is technically infeasible, see Total Removal with HSRS
Sustainable	Flushing	228,657,260.45
Sustainable	HSRS	231,507,568.02
Sustainable	Dredging	229,440,860.27
Sustainable	Trucking	227,412,020.74

Table 5.4 Economic Conclusion of Çubuk Dam I

Strategy yielding the highest aggregate net benefit:	Sustainable
Technique yielding the highest aggregate net benefit:	HSRS
The highest aggregate net benefit is: $	2.315E+08

Table 5.5 Nonsustainable (Decommissioning) Alternatives Details of Çubuk Dam-I

# of years until Partial Removal Option with HSRS is practiced:	Not applicable	years
# of years until retirement for Decommission – with No Removal Option:	85	years
# of years until retirement for Decommission: Partial Removal Option with HSRS:	Not applicable	years
Remaining reservoir capacity at retirement for Decommission – with No Removal Option:	5,068	m³
Remaining reservoir capacity at retirement for Decommission: Partial Removal Option with HSRS:	Not applicable	m³

Table 5.6 Annual Retirement Funds for Decommissioning for Nonsustainable (Decommissioning) Alternatives of Çubuk Dam-I

Annual Retirement Fund Payment for nonsustainable options: Decommission	136	$
Annual Retirement Fund Payment for nonsustainable options: Partial Removal with HSRS	Not applicable	$

Table 5.7 Nonsustainable (Run-off River) Alternatives Details of Çubuk Dam-I

# of years until Partial Removal Option with HSRS is practiced:	Not applicable	years
Approximate # of years until dam is silted for Run-off River – with No Removal Option:	86	years
Approximate # of years until dam is silted for Run-off River – with Partial Removal Option:	Not applicable	years

Table 5.8 Long Term Capacities of Sustainable Alternatives of Çubuk Dam-I

Long term reservoir capacity for Flushing	2,957,561	m³
Long term reservoir capacity for HSRS	3,550,000	m³
Long term reservoir capacity for Dredging	2,879,843	m³
Long term reservoir capacity for Trucking	5,200,929	m³

Table 5.9 Phase-I Lengths of Sustainable Alternatives of Çubuk Dam-I

Approximate # of years until dam is sustained at long term capacity for Flushing	19	years
Approximate # of years until dam is sustained at long term capacity for HSRS	1	years
Approximate # of years until dam is sustained at long term capacity for Dredging	17	years
Approximate # of years until dam is sustained at long term capacity for Trucking	17	years

Table 5.10 Times of Flushing Event in Phase-I

Approximate # of Flushing events until dam is sustained at long term capacity	0	times

Table 5.11 Removal Frequencies for Çubuk Dam-1

Strategy	Technique	Cycle/Phase	Frequency of Removal (years)
Nonsustainable – with Partial Removal	HSRS	Annual cycle	Not applicable
Run-off River (Nonsustainable) – with Partial Removal	HSRS	Annual cycle	Not applicable
Sustainable	Flushing	Phase I	No Flushing occurs
Sustainable	Flushing	Phase II	2
Sustainable	HSRS	Annual cycle	1
Sustainable	Dredging	Phase I	17
Sustainable	Dredging	Phase II	1
Sustainable	Trucking	Phase I	17
Sustainable	Trucking	Phase II	56

Table 5.12 Sediment Removed per event for Çubuk Dam-1

Strategy	Technique	Cycle/Phase	Sediment Removed (m³)
Nonsustainable – with Partial Removal*	HSRS	Annual cycle	Not applicable
Run-off River (Nonsustainable) – with Partial Removal	HSRS	Annual cycle	Not applicable
Sustainable	Flushing	Phase I	0
Sustainable	Flushing	Phase II	84,403
Sustainable	HSRS	Annual cycle	42,202
Sustainable	Dredging	Phase I	N/A
Sustainable	Dredging	Phase II	42,202
Sustainable	Trucking	Phase I	N/A
Sustainable	Trucking	Phase II	2,363,288

Table 5.13 Optimal values of ASD/AST and CLF/CLD/CLT for Çubuk Dam-1

Technique	ASD/AST (%)	CLF/CLD/CLT
Flushing (Phase I)	N/A	60
Flushing (Phase II)	2	
HSRS	1	50
Dredging (Phase I)	N/A	59
Dredging (Phase II)	1	
Trucking (Phase I)	N/A	59
Trucking (Phase II)	56	

Table 5.14 Technical Comments for Çubuk Dam-1

Average expected concentration of sediment to water flushed per flushing event:	19,417	ppm
Average expected concentration of sediment to water released downstream of dam per hydrosuction event:	478	ppm
Average expected concentration of sediment to water removed from reservoir per dredging event:	300,000	ppm

Note: Because reservoir is dewatered prior to a trucking event and river is diverted during a trucking event, material removed is moist sediment (negligible water).

Table 5.15 Number of Truck Loads* Required to Complete Sustainable
Sediment Trucking Removal Option for Çubuk Dam-I

Truck Model Number	m³/Truck Load	Number of Loads (Phase I)	Number of Loads (Phase II)
769D	16.2	N/A	145,882
771D	18	N/A	131,294
773D	26	N/A	90,896
775D	31	N/A	76,235
777D	42.1	N/A	56,135
785B	57	N/A	41,461
789B	73	N/A	32,374
793C	96	N/A	24,618

*1997. Caterpillar Performance Handbook, Ed. 28. CAT Publication by Caterpillar
Inc., Peoria, Illinois, USA. October 1997.

Table 5.16 Number of Dredges Required to Complete Sustainable Sediment Dredging Removal Option
for Çubuk Dam-I

Volume Removed per Dredge (m³/Dredge)	No. of Dredges (Phase I)	No. of Dredges (Phase II)
11,000,000	N/A	1

Table 5.17 Unit Cost of Sediment Removal for Çubuk Dam-I

	Phase I	Phase II
Unit Cost of Dredging ($/m³)	N/A	15.00
Unit Cost of HSRS ($/m³)	1.77	

Currently, Ankara Municipality is cleaning the Çubuk Dam-I by the trucking
method. However, according to RESCON the total trucking cost is $9,453,152 which
is nearly half of the total cost of the dam calculated by the program, $17,474,163.

Once again it should be underlined that RESCON is using a life span of 300 years,
thus it is assumed that the structure of the dam will last for 300 years.

5.3.4 Irrigation water supply dam, Ivriz Dam

Past and present situation of Ivriz Dam

Ivriz Dam is located 10 km southeast of Ereğli, Konya (Figure 5.15). The main purpose
of the dam is irrigation and flood control. Construction of the dam was completed in
1993. The dam has a big siltation problem. The area surrounding the reservoir has no
green cover. Research has been carried out by Sönmez and Dinçsoy (2002) present-
ing possible methods to prevent sediment inflow, and their cost, for Ivriz dam. In this
work sediment inflow calculations have been made using GIS (Geographic Information
System) technology and USLE (Universal Soil Loss Equation). There is no sediment

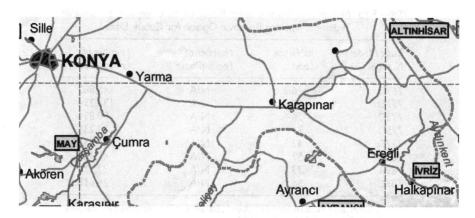

Figure 5.15   Location of Ivriz Dam

Table 5.18 Ivriz Dam Input Data

Parameter	Unit	Value	Source
Reservoir Geometry			
S_0	m³	80,000,000	DSI Web Page (2005)
S_e	m³	75,122,000	Sönmez and Dinçsoy (2002)
W_{bot}	m	75.0	Measured from Drawings (Dams in Turkey, 1991)
SS_{res}		2.0	Measured from Drawings (Dams in Turkey, 1991)
El_{max}	m	1155.0	DSI Web Page (2005)
El_{min}	m	1114.8	DSI Web Page (2005)
El_f	m	1121	Assumed due to not knowing bottom outlet sill elevation
L	m	32,000	Sönmez and Dinçsoy (2002)
Water Characteristics			
V_{in}	m³	104,000,000	Sönmez and Dinçsoy (2002)
Sediment Characteristics			
M_{in}	metric tonnes	340 200	Sönmez and Dinçsoy (2002)
Removal Parameters			
Q_f	m³/s	55	Sönmez and Dinçsoy (2002)
Economic Parameters			
r	decimal	0.08	Koyuncu (2005)
Mr	decimal	0.03	Koyuncu (2005)
PI	$/m³	0.35	Koyuncu (2005)
omc	$/m³	0.10	Koyuncu (2005)
CD	$/m³	3.00	Koyuncu (2005)
CT	$/m³	2.62	Koyuncu (2005)

measurement done by State Hydraulic Works and other governmental or private institutions. The capacity of the dam is 80 hm³ and the height from river bed is 65 m.

RESCON user inputs and results of Ivriz Dam

The RESCON optimization results, using Technical and Economical Parameters of Ivriz Dam, are given below.

Table 5.19 Economic Results for Ivriz Dam

Possible Strategies	Technique	Aggregate Net Present Value
Do nothing	N/A	43,347,725
Nonsustainable (Decommissioning) with Partial Removal	HSRS	43,333,978
Nonsustainable (Run-off River) with No Removal	N/A	43,347,725
Nonsustainable (Run-off River) with Partial Removal	HSRS	43,333,978
Sustainable	Flushing	43,347,011
Sustainable	HSRS	Total Removal with HSRS is technically infeasible, see Partial Removal with HSRS
Sustainable	Dredging	41,485,803
Sustainable	Trucking	38,115,009

Table 5.20 Economic Conclusion for Ivriz Dam

Strategy yielding the highest aggregate net benefit:	Do nothing
Technique yielding the highest aggregate net benefit:	N/A
The highest aggregate net benefit is: $	4.335E+07

Table 5.21 Nonsustainable (Decommission) Alternatives: Details of Ivriz Dam

# of years until Partial Removal Option with HSRS is practiced:	322	years
# of years until retirement for Decommission – with no Removal Option:	324	years
# of years until retirement for Decommission: Partial Removal Option with HSRS:	323	years
Remaining reservoir capacity at retirement for Decommission – with No Removal Option:	907,001 m³	
Remaining reservoir capacity at retirement for Decommission: Partial Removal Option with HSRS:	910,600 m³	

Table 5.22 Annual Retirement Funds for Decommissioning for Nonsustainable (Decommissioning) Alternatives of Ivriz Dam

Annual Retirement Fund Payment for nonsustainable options: Decommission	0	$
Annual Retirement Fund Payment for nonsustainable options: Partial Removal with HSRS	0	$

Table 5.23 Nonsustainable (Run-off River) Alternatives Details of Ivriz Dam-I

# of years until Partial Removal Option with HSRS is practiced:	326	years
Approximate # of years until dam is silted for Run-off River – with No Removal Option:	328	years
Approximate # of years until dam is silted for Run-off River – with Partial Removal Option:	327	years

Table 5.24 Long Term Capacities of Sustainable Alternatives of Ivriz Dam

Long term reservoir capacity for Flushing	47,065,739	m^3
Long term reservoir capacity for HSRS	Not applicable	m^3
Long term reservoir capacity for Dredging	73,653,030	m^3
Long term reservoir capacity for Trucking	78,535,314	m^3

Table 5.25 Phase-I Lengths of Sustainable Alternatives of Ivriz Dam

Approximate # of years until dam is sustained at long term capacity for Flushing	137	years
Approximate # of years until dam is sustained at long term capacity for HSRS	Not applicable	
Approximate # of years until dam is sustained at long term capacity for Dredging	26	years
Approximate # of years until dam is sustained at long term capacity for Trucking	26	years

Table 5.26 Times of Flushing Event in Phase-I, Ivriz Dam

Approximate # of Flushing events until dam is sustained at long term capacity	0 times

Table 5.27 Removal Frequencies for Ivriz Dam

Strategy	Technique	Cycle/Phase	Frequency of Removal (years)
Nonsustainable – with Partial Removal	HSRS	Annual cycle	1
Run-off River (Nonsustainable) – with Partial Removal	HSRS	Annual cycle	1
Sustainable	Flushing	Phase I	No Flushing occurs
Sustainable	Flushing	Phase II	1
Sustainable	HSRS	Annual cycle	Not applicable
Sustainable	Dredging	Phase I	26
Sustainable	Dredging	Phase II	1
Sustainable	Trucking	Phase I	26
Sustainable	Trucking	Phase II	21

Table 5.28 Sediment Removed per event for Ivriz Dam

Strategy	Technique	Cycle/Phase	Sediment Removed (m^3)
Nonsustainable – with Partial Removal	HSRS	Annual cycle	3,599
Run-off River (Nonsustainable) – with Partial Removal	HSRS	Annual cycle	3,599
Sustainable	Flushing	Phase I	0
Sustainable	Flushing	Phase II	244,114
Sustainable	HSRS	Annual cycle	Not applicable
Sustainable	Dredging	Phase I	N/A
Sustainable	Dredging	Phase II	244,114
Sustainable	Trucking	Phase I	N/A
Sustainable	Trucking	Phase II	5,126,398

Table 5.29 Optimal values of ASD/AST and CLF/CLD/CLT, Ivriz Dam

Technique	ASD/AST (%)	CLF/CLD/CLT
Flushing (Phase I)	N/A	41
Flushing (Phase II)	I	N/A
HSRS	I	N/A
Dredging (Phase I)	68	8
Dredging (Phase II)	4	
Trucking (Phase I)	N/A	8
Trucking (Phase II)	89	

Table 5.30 Technical Comments for Ivriz Dam

Average expected concentration of sediment to water flushed per flushing event:	19,471	ppm
Average expected concentration of sediment to water released downstream of dam per hydrosuction event:	40	ppm
Average expected concentration of sediment to water removed from reservoir per dredging event:	300,000	ppm

Note: Because reservoir is dewatered prior to a trucking event and river is diverted during a trucking event, material removed is moist sediment (negligible water)

Table 5.31 Number of Truck Loads Required to Complete Sustainable Sediment Trucking Removal Option, Ivriz Dam

Truck Model Number	m^3/Truck Load	Number of Loads (Phase I)	Number of Loads (Phase II)
769D	16.2	N/A	316,444
771D	18.0	N/A	284,800
773D	26.0	N/A	197,169
775D	31.0	N/A	165,368
777D	42.1	N/A	121,767
785B	57.0	N/A	89,937
789B	73.0	N/A	70,225
793C	96.0	N/A	53,400

Table 5.32 Number of Dredges Required to Complete Sustainable Sediment Dredging Removal Option, Ivriz Dam

Volume Removed per Dredge (m^3/Dredge)	No. of Dredges (Phase I)	No. of Dredges (Phase II)
11.000.000	N/A	I

Table 5.33 Unit Cost of Sediment Removal for Ivriz Dam

	Phase I	Phase II
Unit Cost of Dredging ($/$m^3$)	N/A	3.00
Unit Cost of HSRS ($/$m^3$)	46.31	

5.4 CASE STUDIES FOR HYDROPOWER DAMS

5.4.1 General overview of Coruh Basin Project

Coruh River is born within the borders of Turkey, which has 431 km length. Sediment transport capacity of this river is 5.8 million m³ in a year. The Coruh Basin Project is composed of 27 projects on 10 main streams of Coruh River. After the completion of this project 12 billion kWh will be produced per year. This figure is 27% of the total energy produced by hydroelectric power in Turkey.

Borçka Dam is located on the Çoruh River, in Lower Çoruh Basin, which is in the North-eastern Anatolian Region of the Republic of Turkey. It is one of the Çoruh Basin Project Bunch which includes 27 projects on Çoruh Basin.

The Borçka Dam site is nearly 25 km northwest of Artvin City and 2.5 km upstream of Borçka District. In addition, the dam site is nearly 300 m downstream of the intersection of the Çoruh River and the Murgul Creek, one of the major tributaries of the main stream. Borçka Dam is a clay core, earthfill dam with a reservoir capacity of 418.98×10^6 m³.

The purpose of Borçka Dam is hydroelectric power generation. The installed capacity and total annual energy production are 300 MW and 1,039 GWh respectively. The construction of Borçka Dam was started in 1998 and is now completed and the energy production has been started. Main characteristics of Borçka Dam are given in Table 5.34.

Figure 5.16 A view of Borçka Dam

Table 5.34 Main Characteristics of Borçka Dam

Reservoir Data

Maximum flood water level	187.00 m
Maximum operation water level	185.00 m
Minimum operation water level	170.00 m
Thalweg elevation	103.00 m
Total storage capacity	418.95 hm³
Active storage capacity	150.78 hm³
Dead storage capacity	268.17 hm³
Reservoir maximum surface area (at elev.185)	10.84 km²
Reservoir length	30.50 km

Dam characteristics

Type	Zoned fill with central core
Crest elevation	189.00 m
Height above thalweg	86.00 m
Embankment crest length	557.00 m
Crest Length including concrete structures	728.00 m
Crest width	10.00 m
Total embankment volume	7,785,000 m³

Diversion Facilities

Number of diversion tunnels	2
Cross section type	Horseshoe
Inside diameter	7.50 m
Length (No. 1)	355.00 m
Length (No. 2)	351.00 m
Diversion capacity	1,690 m³/sec
Inlet bottom elevation	104.00 m
Outlet bottom elevation	
Crest elevation of upstream cofferdam	139.00 m
Crest elevation of downstream cofferdam	112.50 m

Spillway

Type	Overflow-controlled
Type of energy dissipation	Chute ending with stilling basin
Sill elevation	168.00 m
Design discharge	10.639 m³/sec
Reservoir elev. at design discharge	187.00 m
Number of gate	4
Type of gates	Tainter gates
Dimension of gates (V/H)	17.00 m/16.00 m

Bottom Outlet

Location	2. Diversion tunnel
Number of bottom outlet	1
Maximum capacity	287.0 m³/sec
Type of bottom outlet gates	Slide gates
Dimension of gates (V/H)	3.50 m/2.50 m
Axis elevation	103.58 m
Inlet sill elevation	140.00 m

Energy Generation Structures
Water Intake Structure

Type	Concrete Gravity
Number	2

(Continued)

Table 5.34 Continued

Approach channel elevation	149.00 m
Water intake axis elevation	154.70 m
Intake invert elevation	151.20 m
Service gate type	Slide gate
Dimensions of service gate (V/H)	5.05 m/7.00 m
Crest Elevation	189.00 m
Reservoir maximum operating level	185.00 m
Reservoir minimum operating level	170.00 m
Penstock	
Type	Partly exposed
Number	2
Inner diameter	7.00 m
Length from service gate up to butterfly valve	207.10 m
Switchyard	
Type	Outdoor
Inlet line voltage and number	380 kV, 2
Outlet line voltage and number	380/154 kV, 2/3
Type of auto-transformer	Single phase
Voltage ratio of auto-transformer	380/154 kV/kV
Number of transformer	6
Transformer capacity	252 (3 × 84) MVA
Powerhouse	
Type	Indoor
Number of units	2
Continuous power	68.40 MW
Installed capacity	300 MW
Continuous energy	600 GWh
Secondary energy	439 GWh
Total Energy	1,039 GWh
Load factor	0.40
Type of inlet valve	Butterfly valve
Inner diameter of inlet valve	5.20 m
Turbine type	Vertical shaft Francis
Turbine axis elevation	92.00 m
Total head	89.00 m
Net head (at operation of one unit)	87.46 m
Maximum discharge	2 × 234.5 m^3/sec
Velocity	136.36 rev/min
Crane capacity of transfer building	100/10 ton
Generator type	Vertical shaft synchronous
Voltage between phases	14.40 kV
Frequency	50 Hz

RESCON user inputs and results of Borcka Dam

RESCON optimization results, using Technical and Economical Parameters of Borçka Dam, are shown below.

Table 5.35 User Input Pages of Borcka Dam

Parameter	Units	Value
Reservoir Geometry		
S_0	(m^3)	418,980,000
S_e	(m^3)	418,980,000
W_{bot}	(m)	385.0
SS_{res}	–	1.0
EL_{max}	(m)	187
EL_{min}	(m)	103
EL_f	(m)	113
L	(m)	30,500
h	(m)	84.0
Water Characteristics		
V_{in}	(m^3)	5,660,000,000
C_v	(m^3)	0.1
T	(°C)	10
Sediment Characteristics		
ρ_d	(tones/m^3)	1.20
M_{in}	(metric tonnes)	10,501,677
Ψ	–	650
Brune Curve No	–	2
Ans	–	3
Type	–	1
Removal Parameters		
HP	–	1
Q_f	(m^3/s)	287
T_f	(days)	5
N	(years)	1
D	(feet)	4
NP	–	1
YA	–	0.3
CLF	(%)	100
CLH	(%)	100
CLD	(%)	100
CLT	(%)	100
ASD	(%)	100
AST	(%)	100
MD	(m^3)	1,000,000
MT	(m^3)	500,000
C_w	(%)	30
Economic Parameters		
E	–	1
C	($/m^3)	0.56
C2	($)	233,829,318.28
R	(decimal)	0.1
Mr	(decimal)	0.1
PI	($/m^3)	0.11
V	($)	15,000,000
omc	–	0.01
PH	($/m^3)	0.011
PD	($/m^3)	0.011
CD	($/m^3)	15.0
CT	($/m^3)	4.0
Flushing Benefits Parameters		
s1	(decimal)	0.9
s2	(decimal)	0.9
Capital Investment		
FI	($)	2,000,000
HI	($)	1,000,000
DU	(years)	25

Table 5.36 Economic Results Summary of Borcka Dam

Possible Strategies	Technique	Aggregate Net Present Value
Do nothing	N/A	6,589,255,222
Nonsustainable (Decommissioning) with Partial Removal	HSRS	6,589,269,107
Nonsustainable (Run-off River) with No Removal	N/A	6,593,100,308
Nonsustainable (Run-off River) with Partial Removal	HSRS	6,593,114,194
Sustainable	Flushing	6,673,099,060
Sustainable	HSRS	Total Removal with HSRS is technically infeasible, see Partial Removal with HSRS
Sustainable	Dredging	6,608,489,562
Sustainable	Trucking	6,630,887,998

Table 5.37 Economic Conclusion of Borcka Dam

Strategy yielding the highest aggregate net benefit:	Sustainable
Technique yielding the highest aggregate net benefit:	Flushing
The highest aggregate net benefit is: $	6.673E+09

Table 5.38 Nonsustainable (Decommissioning) Alternatives: Details of Borcka Dam

# of years until Partial Removal Option with HSRS is practiced:	I	years
# of years until retirement for Decommission – with No Removal Option:	56	years
# of years until retirement for Decommission: Partial Removal Option with HSRS:	56	years
Remaining reservoir capacity at retirement for Decommission – with No Removal Option:	4,888,529	m³
Remaining reservoir capacity at retirement for Decommission: Partial Removal Option with HSRS:	5,382,680	m³

Table 5.39 Annual Retirement Funds for Decommissioning for Nonsustainable (Decommissioning) Alternatives of Borcka Dam

Annual Retirement Fund Payment for nonsustainable options: Decommission	7,248	$
Annual Retirement Fund Payment for nonsustainable options: Partial Removal with HSRS	7,248	$

Table 5.40 Nonsustainable (Run-off River) Alternatives: Details of Borcka Dam

# of years until Partial Removal Option with HSRS is practiced:	I	years
Approximate # of years until dam is silted for Run-off River – with No Removal Option:	57	years
Approximate # of years until dam is silted for Run-off River – with Partial Removal Option:	57	years

Table 5.41 Long Term Capacities of Sustainable Alternatives of Borcka Dam

Long term reservoir capacity for Flushing	209,895,362	m³
Long term reservoir capacity for HSRS	Not applicable	m³
Long term reservoir capacity for Dredging	123,200,378	m³
Long term reservoir capacity for Trucking	411,585,509	m³

Table 5.42 Phase-I Lengths of Sustainable Alternatives of Borcka Dam

Approximate # of years until dam is sustained at long term capacity for Flushing	55	years
Approximate # of years until dam is sustained at long term capacity for HSRS	Not applicable	years
Approximate # of years until dam is sustained at long term capacity for Dredging	40	years
Approximate # of years until dam is sustained at long term capacity for Trucking	38	years

Table 5.43 Times of Flushing Event in Phase-I, Borcka Dam

Approximate # of Flushing events until dam is sustained at long term capacity	9	times

Table 5.44 Removal Frequencies: Borcka Dam

Strategy	Technique	Cycle/Phase	Frequency of Removal (years)
Nonsustainable – with Partial Removal	HSRS	Annual cycle	1
Run-off River (Nonsustainable) – with Partial Removal	HSRS	Annual cycle	1
Sustainable	Flushing	Phase I	5
Sustainable	Flushing	Phase II	2
Sustainable	HSRS	Annual cycle	Not applicable
Sustainable	Dredging	Phase I	40
Sustainable	Dredging	Phase II	1
Sustainable	Trucking	Phase I	38
Sustainable	Trucking	Phase II	38

Table 5.45 Sediment Removed per event for Borcka Dam

Strategy	Technique	Cycle/Phase	Sediment Removed (m³)
Nonsustainable – with Partial Removal*	HSRS	Annual cycle	8,985
Run-off River (Nonsustainable) – with Partial Removal	HSRS	Annual cycle	8,985
Sustainable	Flushing	Phase I	21,330,291
Sustainable	Flushing	Phase II	14,788,981
Sustainable	HSRS	Annual cycle	Not applicable
Sustainable	Dredging	Phase I	N/A
Sustainable	Dredging	Phase II	7,394,491
Sustainable	Trucking	Phase I	N/A
Sustainable	Trucking	Phase II	280,990,641

Table 5.46 Optimal values of ASD/AST and CLF/CLD/CLT for Borcka Dam

Technique	ASD/AST (%)	CLF/CLD/CLT
Flushing (Phase I)	Varies	53
Flushing (Phase II)	7	
HSRS	N/A	N/A
Dredging (Phase I)	N/A	71
Dredging (Phase II)	2	
Trucking (Phase I)	N/A	67
Trucking (Phase II)	78	

Table 5.47 Technical Comments for Borcka Dam

Average expected concentration of sediment to water flushed per flushing event:	186,530	ppm
Average expected concentration of sediment to water released downstream of dam per hydrosuction event:	177	ppm
Average expected concentration of sediment to water removed from reservoir per dredging event:	300,000	ppm

Note: Because reservoir is dewatered prior to a trucking event and river is diverted during a trucking event, material removed is moist sediment (negligible water).

Table 5.48 Number of Truck Loads* Required to Complete Sustainable Sediment Trucking Removal Option for Borcka Dam

Truck Model Number	m^3/Truck Load	Number of Loads (Phase I)	Number of Loads(Phase II)
769D	16.2	N/A	17,345,101
771D	18	N/A	15,610,591
773D	26	N/A	10,807,332
775D	31	N/A	9,064,214
777D	42.1	N/A	6,674,362
785B	57	N/A	4,929,660
789B	73	N/A	3,849,187
793C	96	N/A	2,926,986

*1997. Caterpillar Performance Handbook, Ed. 28. CAT Publication by Caterpillar Inc., Peoria, Illinois, USA. October 1997.

Table 5.49 Number of Dredges Required to Complete Sustainable Sediment Dredging Removal Option for Borcka Dam

Volume Removed per Dredge (m^3/Dredge)	No. of Dredges (Phase I)	No. of Dredges (Phase II)
11,000,000	N/A	1

Table 5.50 Unit Cost of Sediment Removal for Borcka Dam

	Phase I	Phase II
Unit Cost of Dredging ($/$m^3$)	N/A	15.00
Unit Cost of HSRS ($/$m^3$)	4.45	

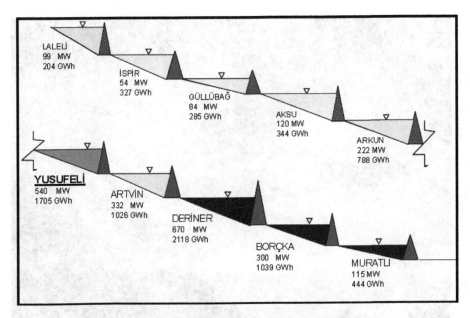

Figure 5.17 Completed and Planned Projects in Çoruh Basin

As a consequence of optimization of Borçka Dam, the highest net aggregate benefit is gained by using the Flushing Method. The highest net aggregate benefit is $4,825,000,000, which is calculated for 300 years perpetuity assumption.

If the characteristics of the Borçka Dam are considered, the flushing method solution is the most appropriate method. Borçka Dam has very high water inflow capacity and it also has a very high sediment inflow capacity. In addition, the elevation of Borçka Dam is significantly high. Because of these characteristics it will be very difficult to remove deposited sediment by using the dredging method or HSRS method. Moreover, the program already warns about dredging, trucking and HSRS. These results actually show that RESCON can be used for Dams in Turkey. Therefore, the most feasible sediment removal system can be decided at the planning stage.

The most feasible method's Flushing characteristics, and all methods analyses, can be followed from Tables 5.36 to 5.50. For example, the long term capacity for flushing is 209,895,362 m^3, which can be seen from Table 5.41. and Phase-I length of flushing and number of flushing events in Phase-I are 55 years and 9 times, respectively.

The sediment removed in Phase-I is 21,330,291 m^3, on the other hand the sediment removed in Phase-II is 14,788,981 m^3. The removed sediment values are different since the long term capacity is smaller than the reservoir capacity.

Although the Borçka Dam has not been designed with the Flushing method in mind, according to the RESCON results Flushing can be used by using the existing sluiceway. Therefore, the Flushing sediment removal operation can be integrated into the existing system of Borçka Dam.

Figure 5.18 A view from Muratlı Dam

As is mentioned, Borçka Dam is one of the completed projects within the scope of the Çoruh Basin Project. The completed and planned projects can be seen in Figure 5.17. In the future, when the planned projects of the Çoruh Basin Project are completed, the total incoming sediment to the Borçka Dam will be decreased. As a result the computed flushing results will change so that the removal of sediment every five years by Flushing may be sufficient. Therefore, when the Çoruh Basin Project is completed, an integrated sediment management should be performed for the entire basin.

5.4.1.1 Muratlı Dam

Muratlı Dam is also one of the Çoruh Basin Projects like Borçka Dam. Muratlı Dam is located on Çoruh River in Lower Çoruh Basin. The dam site is nearly 17 km downstream of Borçka Town, 2 km upstream of Muratlı Town and 44 km northwest of Artvin City.

The purpose of Muratlı Dam is hydroelectric power generation. The installed capacity is 115 MW and the annual energy generation is 444.12 GWh.

The main characteristics of Muratlı Dam are given in Table 5.51 below.

RESCON user inputs and results of Borcka Dam

RESCON optimization results by using Technical and Economical Parameters of Muratlı Dam are shown below.

Table 5.51 Main Characteristics of Muratlı Dam

Reservoir Data

Maximum flood water level	98.00 m
Maximum operation water level	96.00 m
Minimum operation water level	91.00 m
Thalweg elevation	56.00 m
Total storage capacity	74.80 hm^3
Active storage capacity	19.94 hm^3
Dead storage capacity	54.86 hm^3
Reservoir maximum surface area (at elev.185)	4.115 km^2
Reservoir length	18.00 km

Dam characteristics

Type	Rock fill with asphalt lining on the upstream face
Crest elevation	100.00 m
Height above thalweg	44.00 m
Embankment crest length	240.00 m
Crest Length including concrete structures	438.00 m
Crest width	10.00 m
Total embankment volume	1,981,000 m^3

Diversion Facilities

Number of diversion tunnels	2
Cross section type	Horseshoe
Inside diameter	10.00 m
Length (No. 1)	300.65 m
Length (No. 2)	364.00 m
Diversion capacity	1,725.00 m^3/sec
Inlet bottom elevation	59.50 m
Outlet bottom elevation	58.20 m
Crest elevation of upstream cofferdam	72.50 m
Crest elevation of downstream cofferdam	66.00 m

Spillway

Type	Overflow-controlled
Type of energy dissipation	Stilling basin
Sill elevation	79.00 m
Design discharge	10.961 m^3/sec
Reservoir elev. at design discharge	98.00 m
Number of gate	4
Type of gates	Tainter gates
Dimension of gates (V/H)	18.00 m/16.00 m

Bottom Outlet

Location	1. Diversion tunnel
Number of bottom outlet	1
Maximum capacity	250.0 m^3/sec
Type of bottom outlet gates	Slide gates
Dimension of gates (V/H)	3.00 m/2.05 m

Energy Generation Structures
Water Intake Structure

Type	Concrete Gravity
Number	2
Approach channel elevation	75.00 m
Water intake axis elevation	79.75 m

(Continued)

Table 5.51 Continued

Service gate type	Slide gate
Dimensions of service gate (V/H)	5.90 m/7.50 m
Crest Elevation	100.00 m
Reservoir maximum operating level	96.00 m
Reservoir minimum operating level	91.00 m
Penstock	
Type	Partly exposed
Number	2
Inner diameter	7.50 m
Length	90.60 m
Switchyard	
Type	Outdoor
Inlet line voltage and number	154 kV, 2
Outlet line voltage and number	154 kV, 2
Powerhouse	
Type	Indoor
Number of units	2
Continuous power	28.90 MW
Installed capacity	115 MW
Continuous energy	253.34 GWh
Secondary energy	190.78 GWh
Total Energy	444.12 GWh
Load factor	0.44
Type of inlet valve	Butterfly valve
Inner diameter of inlet valve	5.60 m
Turbine type	Vertical shaft Francis
Turbine axis elevation	56.00 m
Total head	37.00 m
Net head (at operation of one unit)	36.04 m
Maximum discharge	180.78 m^3/sec
Velocity	111 rev/min
Generator type	Vertical shaft synchronous
Voltage between phases	13.8 kV
Frequency	50 Hz

Muratlı Dam RESCON Results and Comments

RESCON optimization results, using Technical and Economical Parameters of Muratlı Dam, are shown in Table 5.53.

Similar to the Borçka Dam results, Muratlı Dam RESCON results show that the highest net aggregate benefit is gained by using the Flushing Method. This is an expected result because Borçka Dam and Muratlı Dam are both located on the Çoruh River. In addition, although Muratlı Dam is the last downstream project in Çoruh Basin Project, it is the first constructed one; it is in danger of being filled by sediment.

Similar to the Borçka Dam, the Flushing management technique should be integrated to the existing system of Muratlı Dam. The sedimentation problem in Muratlı Dam could be overcome by Flushing.

Table 5.52 User Input Pages of Muratli Dam

Parameter	Units	Value
Reservoir Geometry		
S_0	(m^3)	74,800,000
S_e	(m^3)	74,800,000
W_{bot}	(m)	385.0
SS_{res}	–	1.0
EL_{max}	(m)	98.0
EL_{min}	(m)	56.0
EL_f	(m)	66
L	(m)	18,000
h	(m)	42.0
Water Characteristics		
V_{in}	(m^3)	6,060,000,000
C_v	(m^3)	0.1
T	$(°C)$	10.0
Sediment Characteristics		
ρ_d	$(tones/m^3)$	1.20
M_{in}	$(metric\ tonnes)$	10,501,677
Ψ	–	650
Brune Curve No	–	2
Ans	–	3
Type	–	1
Removal Parameters		
HP	–	1
Q_f	(m^3/s)	250
T_f	$(days)$	5
N	$(years)$	1
D	$(feet)$	4
NP	–	1
YA	–	0.3
CLF	$(\%)$	100
CLH	$(\%)$	100
CLD	$(\%)$	100
CLT	$(\%)$	100
ASD	$(\%)$	100
AST	$(\%)$	100
MD	(m^3)	1,000,000
MT	(m^3)	500,000
C_w	$(\%)$	30
Economic Parameters		
E	–	1
C	$(\$/m^3)$	0.54
C2	$(\$)$	225,974,930.06
R	$(decimal)$	0.1
Mr	$(decimal)$	0.1
PI	$(\$/m^3)$	0.15
V	$(\$)$	10,644,365.54
omc	–	0.01
PH	$(\$/m^3)$	0.015
PD	$(\$/m^3)$	0.015
CD	$(\$/m^3)$	15.00
CT	$(\$/m^3)$	4.00
Flushing Benefits Parameters		
s1	$(decimal)$	0.9
s2	$(decimal)$	0.9
Capital Investment		
FI	$(\$)$	2,000,000
HI	$(\$)$	1,000,000
DU	$(years)$	25.0

Table 5.53 Economic Results Summary of Muratli Dam

Possible Strategies	Technique	Aggregate Net Present Value
Do nothing	N/A	2,624,004,134
Nonsustainable (Decommissioning) with Partial Removal	HSRS	2,625,065,040
Nonsustainable (Run-off River) with No Removal	N/A	2,780,808,924
Nonsustainable (Run-off River) with Partial Removal	HSRS	2,781,869,830
Sustainable	Flushing	3,673,119,588
Sustainable	HSRS	Total Removal with HSRS is technically infeasible, See Partial Removal with HSRS
Sustainable	Dredging	3,624,713,276
Sustainable	Trucking	3,371,593,293

Table 5.54 Economic Conclusion of Muratli Dam

Strategy yielding the highest aggregate net benefit:	Sustainable
Technique yielding the highest aggregate net benefit:	Flushing
The highest aggregate net benefit is: $	3.673E+09

Table 5.55 Nonsustainable (Decommissioning) Alternatives: Details of Muratli Dam

# of years until Partial Removal Option with HSRS is practiced:	1	years
# of years until retirement for Decommission – with No Removal Option:	18	years
# of years until retirement for Decommission: Partial Removal Option with HSRS:	18	years
Remaining reservoir capacity at retirement for Decommission – with No Removal Option:	458,150	m^3
Remaining reservoir capacity at retirement for Decommission: Partial Removal Option with HSRS:	556,041	m^3

Table 5.56 Annual Retirement Funds for Decommissioning for Nonsustainable (Decommissioning) Alternatives of Muratli Dam

Annual Retirement Fund Payment for nonsustainable options: Decommission	233,433	$
Annual Retirement Fund Payment for nonsustainable options: Partial Removal with HSRS	233,433	$

Table 5.57 Nonsustainable (Run-off River) Alternatives: Details of Muratli Dam

# of years until Partial Removal Option with HSRS is practiced:	1	years
Approximate # of years until dam is silted for Run-off River – with No Removal Option:	19	years
Approximate # of years until dam is silted for Run-off River – with Partial Removal Option:	19	years

Table 5.58 Long Term Capacities of Sustainable Alternatives of Muratli Dam

Long term reservoir capacity for Flushing	29,889,464	m³
Long term reservoir capacity for HSRS	Not applicable	m³
Long term reservoir capacity for Dredging	70,669,897	m³
Long term reservoir capacity for Trucking	70,669,897	m³

Table 5.59 Phase-I Lengths of Sustainable Alternatives of Muratli Dam

Approximate # of years until dam is sustained at long term capacity for Flushing	54	years
Approximate # of years until dam is sustained at long term capacity for HSRS	Not applicable	years
Approximate # of years until dam is sustained at long term capacity for Dredging	1	years
Approximate # of years until dam is sustained at long term capacity for Trucking	9	years

Table 5.60 Times of Flushing Event in Phase-I

Approximate # of Flushing events until dam is sustained at long term capacity	13	times

Table 5.61 Removal Frequencies: Muratli Dam

Strategy	Technique	Cycle/Phase	Frequency of Removal (years)
Nonsustainable – with Partial Removal	HSRS	Annual cycle	1
Run-off river (Nonsustainable) – with Partial Removal	HSRS	Annual cycle	1
Sustainable	Flushing	Phase I	3
Sustainable	Flushing	Phase II	2
Sustainable	HSRS	Annual cycle	Not applicable
Sustainable	Dredging	Phase I	1
Sustainable	Dredging	Phase II	1
Sustainable	Trucking	Phase I	9
Sustainable	Trucking	Phase II	9

Table 5.62 Sediment Removed per event for Muratli Dam

Strategy	Technique	Cycle/Phase	Sediment Removed (m³)
Nonsustainable – with Partial Removal*	HSRS	Annual cycle	5,758
Run-off river (Nonsustainable) – with Partial Removal	HSRS	Annual cycle	5,758
Sustainable	Flushing	Phase I	13,656,729
Sustainable	Flushing	Phase II	8,260,206
Sustainable	HSRS	Annual cycle	Not applicable
Sustainable	Dredging	Phase I	N/A
Sustainable	Dredging	Phase II	4,130,103
Sustainable	Trucking	Phase I	N/A
Sustainable	Trucking	Phase II	37,170,925

Table 5.63 Optimal values of ASD/AST and CLF/CLD/CLT for Muratli Dam

Technique	ASD/AST (%)	CLF/CLD/CLT
Flushing (Phase I)	Varies	71
Flushing (Phase II)	16	
HSRS	N/A	N/A
Dredging (Phase I)	N/A	6
Dredging (Phase II)	100	
Trucking (Phase I)	N/A	50
Trucking (Phase II)	100	

Table 5.64 Technical Comments for Muratli Dam

Average expected concentration of sediment to water flushed per flushing event:	123,226	ppm
Average expected concentration of sediment to water released downstream of dam per hydrosuction event:	124	ppm
Average expected concentration of sediment to water removed from reservoir per dredging event:	300,000	ppm

Note: Because reservoir is dewatered prior to a trucking event and river is diverted during a trucking event, material removed is moist sediment (negligible water).

Table 5.65 Number of Truck Loads Required to Complete Sustainable Sediment Trucking Removal Option for Muratli Dam

Truck Model Number	m^3/Truck Load	Number of Loads (Phase I)	Number of Loads (Phase II)
769D	16.2	N/A	2,294,502
771D	18	N/A	2,065,051
773D	26	N/A	1,429,651
775D	31	N/A	1,199,062
777D	42.1	N/A	882,920
785B	57	N/A	652,121
789B	73	N/A	509,191
793C	96	N/A	387,197

*1997. Caterpillar Performance Handbook, Ed. 28. CAT Publication by Caterpillar Inc., Peoria, Illinois, USA. October 1997.

Table 5.66 Number of Dredges Required to Complete Sustainable Sediment Dredging Removal Option for Muratli Dam

Volume Removed per Dredge (m^3/Dredge)	No. of Dredges (Phase I)	No. of Dredges (Phase II)
11,000,000	N/A	1

Table 5.67 Unit Cost of Sediment Removal for Muratli Dam

	Phase I	Phase II
Unit Cost of Dredging ($/m³)	N/A	15.00
Unit Cost of HSRS ($/m³)	6.95	

Chapter 6

Sensitivity analysis of RESCON

In order to distinguish the most critical input values of RESCON, some parameters are changed and then the program run again in many different instances. The Borçka Dam Case study is used for sensitivity analyses. The basic parameters used in these sensitivity analyses are:

- Unit Benefit of Reservoir: P1
- Discount Rate: r
- Market Interest Rate: Mr
- Salvage Value: V

6.1 UNIT VALUE OF THE RESERVOIR YIELD

In order to see how the results are affected by the changing of user input parameters, some of the important parameters are analyzed. One of them is the Unit value of the reservoir yield, which directly affects the Aggregate Net Benefit.

The Unit Reservoir Yield of Borçka Dam is 0.16 $/m^3 \pm 5%$. The change of the Aggregate Net Benefit is shown in Table 6.1.

The results show that the estimation of the water yield is an important parameter in the calculated aggregate benefit. As the unit value of reservoir yield increases, the total net benefit increases, too.

According to Turkish law, water is the property of the public; it is not a tradable good. However, the price of the service can be changed. Therefore, estimation of the water value is a quite speculative topic. Here we estimate the price of water according to the EPDK. On the other hand, for Çubuk Dam-I, the price of water was estimated according to the available application of the city of the Ankara. Even the water can be served free of charge. Thus the NPV values should not be taken as a profit but as the burden to the state.

However, if one compares different management techniques the program results can be safely used.

Moreover, the changes of unit value of reservoir yield may affect the parameters below:

- Long term reservoir capacity for dredging.
- Long term reservoir capacity for trucking.

Table 6.1 Results of Unit Value of Reservoir Yield Change from 0.11 $/m³ to 0.22 $/m³

Possible Strategies	Technique	Aggregate Net Present Value ($)					
	Water Yield ($/m³)	0.11	0.13	0.15	0.18	0.20	0.22
Do nothing	N/A	4.76352E+09	5.67639E+09	6.58926E+09	7.95856E+09	8.87142E+09	9.78429E+09
Nonsustainable (Decommissioning) with Partial Removal	HSRS	4.76342E+09	5.67635E+09	6.58927E+09	7.95865E+09	8.87158E+09	9.78450E+09
Nonsustainable (Run-of-River) with No Removal	N/A	4.76633E+09	5.67971E+09	6.59310E+09	7.96318E+09	8.87656E+09	9.78995E+09
Nonsustainable (Run-of-River) with Partial Removal	HSRS	4.76623E+09	5.67967E+09	6.59311E+09	7.96328E+09	8.87672E+09	9.79016E+09
Sustainable	Flushing	4.82466E+09	5.74888E+09	6.6710E+09	8.05943E+09	8.98364E+09	9.90786E+09
Sustainable	HSRS	Total Removal with HSRS is technically infeasible, See Partial Removal with HSRS	Total Removal with HSRS is technically infeasible, See Partial Removal with HSRS	Total Removal with HSRS is technically infeasible, See Partial Removal with HSRS	Total Removal with HSRS is tehnically infeasible, See Partial Removal with HSRS	Total Removal with HSRS is technically infeasible, See Partial Removal with HSRS	Total Removal with HSRS is technically infeasible, See Partial Removal with HSRS
Sustainable	Dredging	4.77203E+09	5.68996E+09	6.60849E+09	7.98734E+09	8.90722E+09	9.82764E+09
Sustainable	Trucking	4.78819E+09	5.70916E+09	6.63089E+09	8.01437E+09	8.93704E+09	9.85992E+09

- Approximate # of years until dam is sustained at long term capacity for dredging.
- Approximate # of years until dam is sustained at long term capacity for trucking.
- Removal frequencies of dredging.
- Removal frequencies of trucking.
- Sediment removed per event of trucking.
- Optimal values of ASD/AST and CLF/CLD/CLT of dredging.
- Optimal values of ASD/AST and CLF/CLD/CLT of trucking.
- Number of truck loads required to complete sustainable sediment trucking removal option.

These parameters are computed by RESCON; changes in these parameters are negligible and do not change the suggested best method.

6.2 DISCOUNT RATE

The original value of the discount rate used in the Borçka Dam Case Study is 0.10. However, different values are studied, which are from 0.0000001 to 0.12, in order to see how the results change. The range of the discount rate is very high since aggregate net benefit is very sensitive to the discount rate.

Discount rate affects the aggregate net benefit results but it does not change the best solution. If the discount rate decreases the aggregate net benefit will increase drastically and therefore the differences between the methods can be noticed but the best solution does not change. On the other hand, the discount rate is a different specialty and should be evaluated by experts in economics. It should be noted that the net aggregate benefits should not be interpreted as the profit of the investor.

6.3 MARKET INTEREST RATE

In the economic optimization framework of RESCON, the market interest rate is used only for the calculation of the retirement fund of non sustainable options. Therefore, any change in the market interest rate does not affect the aggregate net benefits and the best solution. The change of retirement funds is shown in Table 6.3.

According to the sensitivity analyses, if the market interest rate increases, any retirement fund for non sustainable options decreases. For example, in order to save the salvage value, the retirement fund is $4794 and $7248 for 0.11 and 0.10 market interest rate respectively. It is an anticipated result from the economical point of view. Since the present value of money is more valuable if the market interest rate is high.

6.4 SALVAGE VALUE

Salvage Value is an important parameter for non-sustainable alternatives, because the annual retirement fund is directly affected by salvage value. From the sensitivity analyses of salvage value, it is obviously observed that if the salvage value increases the annual retirement fund will increase. In addition there will be some changes in

Table 6.2 Results of Discount Rate Change from 0.0000001 to 0.12

| Possible Strategies | Technique | Aggregate Net Present Value ($) | | | | | |
| | | 0.0000001 | 0.00001 | 0.001 | 0.10 | 0.11 | 0.12 |
	Discount rate						
Do nothing	N/A	3.16470E+10	3.16392E+10	3.08736E+10	6.58926E+09	6.01267E+09	5.52258E+09
Nonsustainable (Decommissioning) with Partial Removal	HSRS	3.16755+10	3.16677E+10	3.09007E+10	6.58927E+09	6.01258E+09	5.52243E+09
Nonsustainable (Run-of-River) with No Removal	N/A	7.84671E+14	7.87370E+12	1.05081E+11	6.59310E+09	6.01478E+09	5.52376E+09
Nonsustainable (Run-of-River) with Partial Removal	HSRS	7.846711E+14	7.87373E+12	1.05108E+11	6.59311E+09	6.01469E+09	5.52360E+09
Sustainable	Flushing	5.88623E+15	5.88744E+13	5.99953E+11	6.67310E+09	6.05994E+09	5.54904E+09
Sustainable	HSRS	Total Removal with HSRS is technically infeasible, See Partial Removal with HSRS	Total Removal with HSRS is technically infeasible, See Partial Removal with HSRS	Total Removal with HSRS is technically infeasible, See Partial Removal with HSRS	Total Removal with HSRS is tehnically infeasible, See Partial Removal with HSRS	Total Removal with HSRS is technically infeasible, See Partial Removal with HSRS	Total Removal with HSRS is technically infeasible, See Partial Removal with HSRS
Sustainable	Dredging	5.95557E+15	5.95555E+13	5.95442E+11	6.60849E+09	6.02272E+09	5.52790E+09
Sustainable	Trucking	6.26783E+15	6.26789E+13	6.27349E+11	6.63089E+09	6.03504E+09	5.53473E+09

Table 6.3 Results of Market Interest Rate Change from 0.08 to 0.13

	Annual Retirement Fund					
Market Interest Rate (decimal)	0.08	0.09	0.10	0.11	0.12	0.13
Possible Strategies						
Do Nothing	16,343	10,913	7,248	4,794	3,161	2,080
Nonsustainable (Decommissioning) with Partial Removal	16,343	10,913	7,248	4,794	3,161	2,080

Table 6.4 Results of Salvage Value Change from $5,000,000 to $30,000,000

	Annual Retirement Fund					
Salvage Value ($)	5,000,000	10,000,000	15,000,000	20,000,000	25,000,000	30,000,000
Possible Strategies						
Do Nothing	2,416	4,832	7,248	9,663	12,079	14,495
Nonsustainable (Decommissioning) with Partial Removal	2,416	4,832	7,248	9,663	12,079	14,495

Aggregate Net Benefit of non-sustainable alternatives but these are negligible. Related sensitivity analyzes results shown in Table 6.4.

6.5 DISCUSSIONS ABOUT RESCON INPUT VALUES

6.5.1 Unit Benefit of Reservoir Yield (P1)

Unit Benefit of Reservoir Yield depends on the purpose of the dam. The dam may be used for the purposes of municipal, irrigation or energy production. Therefore the unit value of the water has shown differences. In addition the country, and even the region, affects the water value. Because of these reasons, P1 should be calculated separately for each dam, and then put in RESCON.

Since Çubuk Dam-1 is used only for domestic purposes, P1 is obtained from ASKİ 06.03.2008 dated Water and Wastewater Tariff.

On the other hand, for hydropower dams it is very difficult to define P1. The P1 value should be put in RESCON as a unit of $/m^3$. However, in practice the P1 value of hydropower dams has a different unit; the market prefers a unit of $/kWh. Therefore, this unit should be converted to unit $/m^3$ for RESCON usage. In this study, according to reservoir inflow and total head the P1 value is calculated for Borçka Dam and Muratlı Dam separately. P1 is taken as 9.67 Ykr/kWh according to the EPDK (Energy Market Regularity Authority). The converting formula for Borçka Dam is given below:

$$\text{Power} = 9.81 \times Q \times h \times n$$

where:
Q: reservoir inflow (m³/s)
n: efficiency
h: total head (operation level – turbine axis elevation)
Turbine axis elevation: 92 m
Maximum operation water level: 185 m
Minimum operation water level: 170 m

The installed capacity of 300 MW is taken as power value. The reservoir inflow is calculated from the mean annual reservoir inflow 5.66×10^9 m³, that Q is 180 m³/s. Efficiency for both turbine and generator, is approximately 0.9. The total head is not a constant value. Although the turbine axis elevation does not change, the operation level changes between 185 m and 170 m. Therefore, the total head changes between 93 m to 78 m. As a consequence, the operation level is taken as 185 m and so that the total head is calculated as 93 m. It is assumed that the dam works under the full performance condition and all flow is used for energy production.

The benefit gained from the generated power $= 300 \times 1000 \times 9.67 = 2,901,000$ Ykr.
This power is generated with a 180 m³/s water flow through 93 m head.

Thus the flow of 180 m³/s for this reservoir makes 2,901,000 Ykr. Then the unit benefit gained from per m³ of water is 19.5 Ykr/m³.

The parity rate is taken as 1.4 according to the DSI unit prices, therefore $P1 \approx 14$ Cent/m³ $= 0.14$ $/m³

Although, the market energy value is taken as 9.67 Ykr according to the EPDK, it is subject to change and volatile. The electric market operation system is shown schematically below; it is a complex system. That is why it is very difficult to come up with an exact energy value; it is an exchange commodity. For example, the energy value of TEDAŞ is 23 Ykr/kWh. However, the independent producer might be agreed with the customer at 21 YKr/kWh. In another case, the firms might produce their own energy and sell the remaining to the TETAŞ and there may be a special contract between them about the selling price. Because of this complexity over the value of EPDK, the most confident value is used.

6.5.2 Total cost of dam construction

According to the economic optimization framework of RESCON, the program automatically calculates the total cost of dam construction, C2. For example, since Borçka Dam reservoir capacity is smaller than 500,000,000 m³ the program automatically calculates unit cost of construction (c) by using the formula:

$$c = 3.5 - 0.53 * \text{LN}(So/1,000,000)$$

$$= 3.5 - 0.53 * \text{LN}(418,980,000/1,000,000) = 0.30 \text{ \$/m}^3$$

Then C2 = 0.30 * 418,980,000 = $125,674,606.
However, the total cost of dam construction is known as $233,829,318.

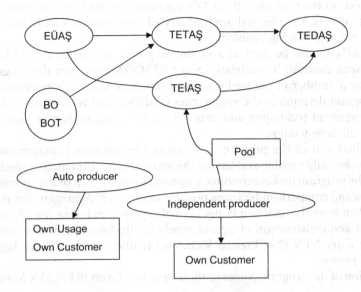

Figure 6.1 Energy Productions and Sale Relation

This study aims to examine the plausibility of the practical usage of RESCON for Turkey's conditions. Coming from this approach, sometimes the necessary modifications have been made in the program. In order to obtain the approximate C2 value for Borçka Dam, the formula has been changed as following:

$$c = 3.5 - 0.487 * LN(So/1,000,000)$$
$$= 3.5 - 0.487 * LN(418,980,000/1,000,000) = 0.56\ \$/m^3$$

Then C2 = 0.56 * 418,980,000 = \$234.452.874.

The formula above is definitely only valid for Borçka Dam. However, if there were a database for the cost of dam construction in Turkey, a general modification could have been done to make a major contribution to RESCON. Therefore, the cost of dam construction in Turkey should be immediately analyzed in further academic works.

6.5.3 Salvage value

Since decommissioning of a dam is not a common execution worldwide the estimation of the salvage value is not exactly defined in the literature. There are some papers and research, but no definite calculation procedure. Therefore, as salvage value the expropriation value is used in this study.

6.6 STRENGTH AND LIMITATIONS OF RESCON

RESCON does not make an analysis about the feasibility of dredging and trucking. It gives some cautions but gives the responsibility of evaluating the outcomes of these two methods to the user since RESCON assumes that these two methods are always feasible. However, the physical applicability of these methods and placement of the removed sediment is a big problem.

RESCON should be used as a preliminary tool. Its results should be evaluated carefully with caution. It is advised by the RESCON team that the program should be used for a number of isolated reservoirs rather than a single reservoir. This excel based program determines the engineering feasibility and economical values of sediment management techniques and ranks them. The program can be used for existing dams as well as new dams.

The final aim of the program is to select the sediment management technique which is technically feasible and having the maximum net benefit. Site specific data are crucial. The program makes economical optimization for each of the sediment removal techniques and comparison becomes possible in this way. Aggregate Net Benefit is the benefit taken from the dam minus any kind of expenses including installation of HSRS equipment and construction of new channels for flushing operations over the entire life of the dam. NPV (Net Present Value) is the discounted value of Aggregate Net Benefit to present.

Solution of the program comes to the user in two forms (RESCON Manual, 2003):

1   SUSTAINABLE, where the reservoir capacity can be maintained at an original or a lower capacity,

2   NON-SUSTAINABLE, where reservoir fills with sediment in a finite time. This solution divides into two:

   a)   The dam is decommissioned at an optimally determined time allowing salvage value (=cost of decommissioning minus any benefits due to decommissioning) to be collected at this time; or

   b)   The dam is maintained as a "run-off river" project, even after the reservoir is silted.

If decommissioning is the best solution an annual retirement fund is calculated by the program. For sustainable solutions NPV is calculated as well as for the run-off river option. This creates a chance to compare the outcomes of each technique.

Environmental results are also important, even if a sediment removal technique leads to a sustainable solution. Since removed sediment is also a big problem for the neighbourhood of the dam or for the next dam.

RESCON is a program to be used for a single isolated reservoir and using RESCON for systems of reservoirs (reservoirs following each other) may not give good results. Since application of flushing or HSRS changes the amount of sediment inflowing to the next reservoir. This lowers the economic life of the dam whose inflowing sediment is higher than before. Therefore, in order for RESCON to be used for systems of reservoirs, modifications should be made to the program code (Palmieri et al., 2003).

References

Ackers, P. and White W.R., 'Sediment transport: New approach and analysis', *Journal of Hydraulics Division*, ASCE, Vol. 98, No. HY11, Proceeding Paper 10167, pp. 2041–2060, 1973.

Altun, Ş., Sarvary, I. and Tiğrek, Ş., 'Prediction of Reservoir Siltation Case Studies for Turkey', *HYDRO 2004*, Porto, Portugal (in CD), 2004.

Amini, A., and Fouladi, C., 'Sediment flushing at the Sefidrud Reservoir. Proceedings of the 2nd International Workshop on Alluvial River Problems', *R.J. Garde, ed.*, University of Roorkee, India. pp. 97–102, 1985.

Annandale, G.W., 'Engineering and Hydrosystems Inc.', Personal Communication, 2008.

ANRE (Agency of Natural Resources and Energy), 'Sedimentation status of reservoirs and ponds for hydropower projects in fiscal year of 1983–2001', *Electric Power Civil Engineering*, No. 192–295 (in Japanese), 1984–2001.

Atkinson, E., 'The Feasibility of Flushing Sediment from Reservoirs' TDR Project R5839, Report OD 137, HR Wallingford, November 1996.

Bagnold, R.A., 'An approach to the sediment transport problem from general physics', U.S. Geological Survey, Professional Paper, 422-J, 1966.

Basson, G., 'Hydraulic Measures to deal with reservoir sedimentation: flood flushing, sluicing and density current venting', In *Proceedings of the 3rd International Conference on River Flood Hydraulics*. University of Stellenbosch, South Africa, 1997

Basson, G., 'Hydropower Dams and Fluvial Morphological Impacts – An African Perspective', *United Nations Symposium on Hydropower and Sustainable Development*, Beijing International Convention Centre, Beijing, China, United Nations Dept. of Economics and Social Affairs 2004: 1–16.

Basson, G. and Rooseboom, A., *Dealing with Reservoir Sedimentation*, Water Research Commission, South Africa, December 1997.

Bishop, A.A., Simons, D.B. and Richardson E.V., 'Total Bed Material Transport', *Journal of the Hydraulics Division*, ASCE, Vol. 91, HY2, 1965.

Borland, W.M. and Miller, C.R., 'Distribution of Sediment in Large Reservoirs', *J. Hydraulics Div. ASCE*, 84 (HY2), 1958.

Borland, W.M. and Miller, C.R., "Reservoir Sedimentation" in *River Mechanics* (editor: H. Shen), Colorado State Univ., Water Resources Publication, Fort Collins, Colo, 1971.

Brown, C.B., 'Sediment transportation', *Engineering Hydraulics*, ed. H. Rouse, New York, 1950.

Brownlie, W.R., 'Prediction of flow depth and sediment discharge in open channels, and Compilation of fluvial channel data: laboratory and field', Rep. No. KH-R-43A&B, W.M. Keck Lab. of Hydr. and Water Resources, California Institute of Technology, Pasadena, California, USA., 1981.

Brune, G.M., 'Trap Efficiency of Reservoirs.' *Trans. Am. Geophysical Union,* 34(3), pp. 407–418, 1953.

'Caterpillar Performance Handbook', Ed. 28. CAT Publication by Caterpillar Inc., Peorian Illinois, USA. October 1997.

Çetinkaya, O.K., 'Management of Reservoir Sedimentation Case Studies from Turkey', M. Sc. Thesis, Middle East Technical University, 2006.

Chang, F.M., Simons, D.B. and Richardson, E.V., 'Total bed material discharge in alluvial channels', *U.S. Geological Survey water supply paper,* 1498-L, 1965.

Chen, Y.H., Lopez, J.L. and Richardson, E.V., 'Mathematical Modeling of Sediment Deposition in Reservoirs, Journal of Hydraulics Division', *Proceedings of the ASCE,* Vol. 104, No. HY12, pp. 1605–1616, 1978.

Chen, J. and Zhao, K., 'Sediment management in Nanqin Reservoir', *International Journal of Sediment Research,* Vol. 7(3), pp. 71–84, 1992.

Colby, B.R., 'Practical computations of bed-material discharge', *Journal of Hydraulics Division,* ASCE, Vol. 90, No. HY2, 1964.

'Dams in Turkey', DSI (State Hydraulic Works), 1991.

Depeweg, H. and Mendez, V.N., A New Approach to Sediment Transport in the Design and Operation of Irrigation Canals, 2007.

Dou, G., 'Similarity theory and its application to the design of total sediment transport model', *Research Bulletin of Nanjing Hydraulic Research Institute,* China, 1974.

Dou, G., 'Total Sediment Transport in Rivers and Its Physical and Mathematical Modeling', *Proceedings of the International Symposium on Sediment Transport Modeling, New Orleans, Louisiana,* Hydraulics Div. of ASCE, pp. 39–44, 1989.

'DSI Report', obtained by personal communication, 2005.

'DSI Web Page', http://www.dsi.gov.tr, 2005.

DuBoys, M.P., 'Le Rhone et les Rivieres a Lit Affouillable', *Annales de Ponts et Chausses,* sec. 5, Vol. 18, pp. 141–195

Durgunoğlu, A., 'An Approach to Estimate Bed Material Load Sedimentation in Reservoirs', Master Thesis, Department of Civil Engineering, METU, Ankara, 1980.

Einstein, H.A., 'Formula for the transportation of bed-load', *Transactions of the ASCE,* Vol. 107, 1942.

Einstein, H.A., 'The bed-load function for sediment transportation in open channel flows', U.S. Department of agriculture soil conservation Service Technical Bulletin no. 1026, 1950.

Engelund, F. and Hansen, E., A monograph on sediment transport in alluvial streams, Teknisk Forlag, Copanhagen, 1972.

Engelund, F. and Fredsøe, J., 'A sediment transport model for straight alluvial channels', *Nordic Hydrology,* Vol. 7, pp. 293–306, 1976.

Exner, F.M., 'Über die Wechselwirkung zwischen Wasser und Geschiebe in Flüssen', *Sitzungberger, Akad. Wissenschaften,* pt. IIa, Bd. 134, Wien, A, 1925.

Fan, J., 'Methods of preserving reservoir capacity', *Methods of Computing Sedimentation in Lakes and Reservoirs: A contribution to the International Hydrological Programme,* IHP-II Project A. 2.6.1 Panel, S. Bruk, ed., Unesco, Paris, 1985

Fan J. and Morris G.L., 'Reservoir Sedimentation I: Delta and Density Current Deposits', *Journal of Hydraulic Engineering,* Vol. 118, No. 3, pp. 354–369, 1992.

Fan, J. and Morris, G.L., 'Reservoir Sedimentation II: Reservoir Desiltation and Long-term Storage Capacity', *Journal of Hydraulic Engineering,* Vol. 118, No. 3, 1992b.

Fan, J., 'An Overview of Preserving Reservoir Storage Capacity', *Proc. 1995 Intl. Workshop on Reservoir Sedimentation at San Francisco,* Federal Regulatory Commission, Washington, D.C., 1995.

Garcia, H.M., 'Sediment Transport and Morphodynamics', Chapter 2, *Sedimentation Engineering: Theories, Measurements, Modeling and Practice: Processes, Management, Modeling, and Practice*, Ed. Marcelo H. Garcia (ASCE Manual and Reports on Engineering Practice No. 110), 2007.

Gill, M.A., 'Nonlinear Solution of Aggradation and Degradation in Channels', *Journal of Hydraulic Research*, Vol. 25, No. 5, 1987.

Göbelez, Ö., 'Experimental Analysis of the Flow through a Bottom Outlet on the Threshold of Motion of Particles', Msc. Thesis. Middle East Technical University, Ankara, 2008.

Graf, H.W., 'Fluvial Hydraulics: Flow and Transport Processes in Channels of Simple Geometry', in collaboration with **M.S. Altınakar**, John Wiley & Sons, 1998 (reprinted in 2001).

Graf, H.W., 'Fluvial Hydraulics', in collaboration with **M.S. Altınakar**, PROGRAMS prepared by Altınakar, M.S, http://lrhwww.epfl.ch/books/books.html, 1998.

Göğüş, M., 'Reservoir Sedimentation, Sediment Transport Technology', pp. 353–377, General Directorate of State Hydraulic Works Technical Research and Quality Control Department, Ankara – Turkey, 2007.

Guy, H.P., Simons, D.B. and Richardson, E.V., 'Summary of alluvial channel data from flume experiments', U.S. Geological Survey professional paper, 462-1, pp. 1965–1961, 1966.

Harris, C.K., 'Sediment Transport Process in Coastal Environments', Lecture Notes, http://www.vims.edu/~ckharris/MS698_03/lecturenotes.html, 2003.

Haested Methods, Dyhouse, G., Hatchett, J. and Benn, J., 'Floodplain Modeling Using HEC-RAS', HAESTED PRESS, Waterbury, CT USA, ISBN: 0-9714141-0-6, 2003.

Henderson, F.M., *Open Channel Flow*, Macmillan Company, New York, USA, 1966.

Hjulström, F., 'Studies of the Morphological Activity of Rivers as Illustrated by the River Fyris', *Bulletin, Geological Institute of Upsala*, Vol. XXV, Upsala, Sweden, 1935.

Holly, F.M. and Rahuel, J.L., 'New Numerical/Physical Framework for Mobile-Bed Modelling', *Journal of Hydraulic Research*, Vol. 28, No. 4, pp. 401–415, 1990.

Hotchkiss, R.H. and Huang, X., 'Hydrosuction Sediment-Removal Systems (HSRS): Principles and Field Test', *Journal of Hydraulic Research*, pp. 479–489, June 1995.

Howard, C.D.D., 'Operations, Monitoring and Decommissioning of Dams', *Thematic Review IV.5 prepared for the World Commission of Dams*, Cape Town, March 2000.

Huang, J.V. and Greimann, B., *User's Manual for SRH1-D (Sedimentation and River Hydraulics – One Dimension) V2.0.5*, USBR, 2007.

Huffaker, R. and Hotchkiss, R., 'Economic Dynamics of Reservoir Sedimentation Management: Optimal Control with Singularly Perturbed Equation of Motion', *Journal of Economic Dynamics & Control* 30 (2006), 2553–2575.

Inglis, C.C., 'Discussion of "Systematic Evaluation of River Regime", by Charles R. Neill and Victor J. Galey', *Journal of the Waterways and Harbors Division*, ASCE, Vol. 94, No. WW1, Proc. Paper 5774, pp. 109–114, 1968.

'In-stream Aggregate Extraction and Reclamation Guidance Document', Prepared for Colorado Department of Natural Resources Division of Minerals and Geology, 1998. http://mining.state.co.us/General%20Reports.htm.

IRTCES, 'Lecture Notes of the Training Course on Reservoir Sedimentation', *International Research and Training Centre on Erosion and Sedimentation*, Tsinghua University, Beijing, China, 1985.

Jaggi, A.L. and Kashyap, B.R., 'Desilting of Baira Reservoir of Baira Siul Project', *Irrigation and Power*, October 1984.

Jarecki, E.A. and Murphy, T.D., 'Sediment Withdrawal Investigation – Guernsey Reservoir', Paper No. 91, *Proc. Inter-Agency Sedimentation Conference*, Misc Publication No. 970, US Dept. of Agriculture, 1963.

Kalinske, A.A., 'Movement of sediment as bed-load in rivers', *Transactions of the American Geophysical Union*, Vol. 28, No. 4, 1947.

Karim, M.F. and Kennedy, J.F., 'Menu of Coupled Velocity and Sediment Discharge Relations for Rivers', *Journal of Hydraulic Engineering*, Vol. 116, No. 8, pp. 978–996, 1990.

Kawashima, S., Johndrow, T.B., Annandale, G.W. and Shah, F., Reservoir Conservation, Volume-II: RESCON Model and User Manual, The World Bank, June 2003.

Kemalli, O., 'Measurements of Velocity Profiles by Using Particle Image Velocimeter', MSc Dissertation, METU, Ankara, 2009.

Kılıç, R., Çubuk I (Ankara) Baraj Gölünde Depolanan Sedimentlerin Sedimentolojisi ve Mineralojisi, Gazi Üniversitesi, Cilt 1, No. 1, 97–112, 1986.

Kılıç, R., Çubuk 1 (Ankara) Barajı'nda Siltasyonun İncelenmesi, Doktora Tezi, İnşaat Mühendisliği Bölümü, Gazi Üniversitesi, Ankara, 1984.

Kostic, S. and Parker, G., 'Progradational Sand-Mud Deltas in Lakes and Reservoirs Part 1: Theory and Numerical Modeling', Journal of Hydraulic Research, Vol. 41, No. 2, pp. 127–140, 2003.

Hu, C., 'Controlling reservoir sedimentation in China', Hydropower & Dams, March 1995, pp. 50–52 , 1995

Lane, E.W., 'Report on Subcommittee on Sediment Transport Terminology', Trans. AGU, Vol. 28, No. 6, 1947.

Laursen, E.M., 'The total sediment load of streams', Journal of Hydraulics Division, ASCE, Vol. 84, No. 1, pp. 1530–1536, 1958.

Liu, J., Liu, B. and Ashida, K., Reservoir Sedimentation Management in Asia, Fifth International Conference on. Hydro-Science and Engineering, ICHE, September 18–21, Warsaw, Poland, pp. 1–10, 2002.

Ludwig, W., 'River Inputs to the Mediterranean Sea', http://www.cnrm.meteo.fr/hymex/global/documents/Presentations_janvier_2007/Session_E/Rivers_input_Ludwig_Wolfgang.pdf, last accessed date May 2011.

Mahmood, K., 'Reservoir Sedimentation: Impact, Extent and Mitigation', World Bank Technical Paper Number 71, Washington, D.C., September 1987.

Meland, N. and Normann, J.O., 'Transport of single particles in bed-load motion', Geografiska Annaler, Vol. 48, A, 1966.

Meyer-Peter, E. and Müller, R., 'Formula for bedload transport', Proceedings of the 2nd Meeting of the IAHR, Stockholm, 39–64, 1948.

Meyer-Peter, E. and Müller, R., 'Formula for Bed-Load Transport', Proceedings of International Association for Hydraulic Research, 2nd Meeting, Stockholm, 1948.

Molinas, A. and Wu, B., 'Comparison of fractional Bed-Material Load Computation Methods in Sand-Bed Channels', Earth Surface Processes and Landforms, Vol. 25, pp. 1045–1068, 2000.

Molinas, A. and Wu, B. S., 'Transport of sediment in large sand bed rivers', J. Hydraul. Res. 39(2), 135–146, 2001.

Molinas, A. and Yang, C. T., 'Computer Program User's Manual for GSTARS', U. S. Bureau of Reclamation Engineering and Research Center, Denver, CO, 80225, 1986.

Morris, G.L. and Fan, J., Reservoir Sedimentation Handbook, McGraw-Hill, New York, 1997.

Palmieri, A., Shah, F., Annandale, G.W. and Dinar, A., Reservoir Conservation, Volume-I: The RESCON Approach, The World Bank, June 2003.

Paphitis, D., 'Sediment movement under unidirectional flows: an assessment of empirical threshold curves', Coastal Engineering 43, 227–45, 2001.

Parhami, F., 'Sediment control methods in Sefid-Rud Reservoir Dam (Iran)', Proceedings of the Third International Symposium on River Sedimentation, the University of Mississippi, March 31–April 4, 1986. S.Y. Wang, H.W. Shen, and L.Z. Ding, eds, pp. 1047–1055, 1986.

Parker, G. and Toniolo, H., '1-D Numerical Modeling of Reservoir Sedimentation', Proceedings of IAHR Symposium on River, Coastal and Estuaries Morphodynamics, Barcelona, Spain, 2003, pp. 457–468.

Parker, G., Klingeman, P.C. and McLean, D.G., 'Bed-load and size distribution in paved gravel-bed streams', *Journal of Hydraulics Division*, ASCE, Vol. 108, No. HY4, pp. 544–571, 1982.

Parker, G., 'Surface-based bed-load transport relationship for gravel rivers', *Journal of Hydraulic Research*, Vol. 28, No. 4, pp. 417–436, 1990.

Partheniades, E., *Handbook of Coastal and Ocean Engineering*, Gulf Publishing Co., Houston, Texas, 1990.

Poulos, S.E. and Collins, M.B., 'A quantitative evaluation of riverine water/sediment fluxes to the Mediterranean basin: natural flows, coastal zone evolution and the role the dam construction', In: S.J. Jones and L.E. Frostick [Eds] *Sediment Flux to Basins: Causes, Controls and Consequences*. Geological Society, London, Special Publications, 227–245, 2002.

Pulcuoğlu, B., 'Comparative Study on Sediment Transport Equations For Delta Formations in Reservoirs', *MSc Dissertation*, METU, Ankara, 2009.

Reid, L.M. and Dunne, T., 'Rapid Evaluation of Sediment Budgets' Catena Verlag, Reiskirchen, 1996.

Rottner, J., 'A formula for bed-load transportation', *La Houille Blanche*, Vol. 14, no. 3, pp. 285–307, 1959.

Rubey, W.W., 'Settling Velocities of Gravel, Sand and Silt Particles', *American Journal of Science*, 1933.

Schmidt, G., 'Siltation Experience on Roseires Dam', *Coyne et Bellier*, Paris, 1983.

Schoklitsch, A., 'Handbuch des wasserbaues', *Springer-Verlag*, New York, 1950.

Shammaa, Y., Zhu, D.Z. and Rajaratnam, N., 'Flow Upstream of Orifices and Sluice Gates', *Journal of Hydraulic Engineering*, Vol. 131, No. 2, February 1, 2005.

Shen, H.W. and Hung, C.S., 'An engineering approach to the total bed material load by regression analysis', *Proceedings of the sedimentation symposium*, Chap. 14, pp. 14-1, 14-17 (1972).

Shields, I.A., 'Application of Similarity Principles and Turbulence Research to Bed-Load Movement', Hydrodynamics Laboratory of California Institute of Technology, Publication No. 126, Pasedena, California, 1936 (Translated by Ott, W.P., and van Uchelen, J.C.).

Simons, D.B. and Şentürk, F., *Sediment Transport Technology*, Water Resources Publications, Fort Collins, Colorado, 1977 (Reprinted 1992).

Sönmez, B. and Dinçsoy, Y., 'Konya-Ereğli İvriz Barajı Erozyon ve Rüsubat Kontrolu Planlama Raporu', State Hydraulic Works, Ankara, 2002.

Straub, L.G., 'Missouri River Report', In-House Document 238, 73rd Congress, 2nd Session, U.S. Government Printing Office, Washington D.C., p.1135.

Tarela, P.A. and Menendez, A.N., A Model to Predict Reservoir Sedimentation, Lakes and Reservoirs: Research and Management, Vol. 4, pp. 121–133, 1999.

Tarela, P.A., 'Modelacion Matematica del Fenomeno de Sedimentacion en Embalses', Fellowship Report, National Council of Scientific Research (CONICET), Argentina, 1995.

Tingsanchali, T. and Supharatid, S., 'Experimental Investigation and Analysis of HEC-6 River Morphological Model', *Hydrological Processes*, Vol. 10, pp. 747–761, 1996.

Toffaleti, F.B., 'Definitive computations sand discharge in rivers', *Journal of Hydraulics Division*, ASCE, Vol. 95, No. HY1, pp. 225–246, 1969.

Toniolo, H. and Parker, G., 1D 'Numerical modeling of reservoir sedimentation', *Proceedings of the IAHR Symposium on River, Coastal and Estuarine Morphodynamics*, Barcelona, Spain, pp. 457–468, 2003.

Turner, T.M., *Fundamentals of Hydraulic Engineering*, American Society of Civil Engineers, 2nd Ed., New York, 1996.

USACE-1977, 'Scour and Deposition in Rivers and Reservoirs HEC-6 *User's Manual Draft*, US Army Corps of Engineers Hydrologic Engineering Center. USA, 1977.

USACE-1993, 'Status and New Capabilities of Computer Program HEC-6: Scour and Deposition in Rivers and Reservoirs. US Army Corps of Engineers Hydrologic Engineering Center. USA, 1993.

Vanoni, V.A., 'Sedimentation Engineering', ASCE Task Committee for the Preparation of the Manual on Sedimentation of the Sedimentation Committee of the Hydraulics Division, 1975 (Reprinted 1977).

Van Rijn, L.C., 'Sediment transport, part I: bed load transport', *Journal of Hydraulic Engineering*, ASCE, Vol. 110(10), pp. 1431–1456, 1984.

Van Rijn, L.C., 'Mathematical model of morphological processes in the case of suspended sediment transport', PhD thesis, Delft University of Technology, Delft, The Netherlands, 1987.

Velikanov, M.A., 'Gravitational theory of sediment transport', *Journal of Science of Soviet Union, Geophysics*, Vol. 4, 1954.

Vituki, 'Application of Hydrology in Water and Environmental Management', *Lecture Notes Post-Graduate Course on Hydrology*, Budapest, HUNGARY, 2003.

Wang, W-C., 'Applicability of reservoir desilting/dredging methods of Mainland China to reservoirs in Taiwan', *Sinotech Engineering Consultants*, Inc. pp. 54–57, 1996.

White, W.R., 'Contributing Paper: Flushing of Sediments from Reservoirs', prepared for *Thematic Review IV. 5: Operations, Monitoring and Decommissioning of Dams*, HR Wallingford, UK, 2000.

White, R., 'Evacuation of Sediments from Reservoirs', *Thomas Telford Publishing*, London. ISBN: 0 7277 2953 5, 2001.

Williams, S.G. and Hazen, A., 'Hydraulic Tables, The Elements of Gagings and the Friction of Water Flowing in Pipes Aqueducts, Sewers, etc. As Determined By The Hazen and Williams Formula and the Flow of Water Over Sharp-Edged and Irregular Weirs, and The Quantity Discharged as Determined by Bazin's Formula and Experimental Investigations Upon Large Models', *second edition, revised and enlarged second, thousand*, John Wiley & Sons, New York, 1911.

Wilcock P.R. and Crowe J.C., 'Surface-based Transport Model for Mixed-Size Sediment', *Journal of Hydraulic Engineering*, Vol. 129, No. 2, pp. 120–128, 2003.

Wu, W., Wang, S.S.Y. and Jia, Y., 'Nonuniform Sediment Transport in Alluvial Rivers', *Journal of Hydraulic Research*, Vol. 38, No. 6, pp. 427–434, 2000.

Wu, W., Computational River Dynamics, Taylor & Francis, Leiden, Netherlands, 2007.

Yalın, M.S., 'An expression for bed-load transportation', *Proceedings of ASCE*, Vol. 89, HY3, 1963.

Yalın, M.S., Mechanics of Sediment Transport, Pergamon Press, 1972.

Yang S-Q. and Lim S-Y., 'Total Load Transport Formula for Flow in Alluvial Channels', *Journal of Hydraulic Engineering*, Vol. 129, No. 1, pp. 68–72, 2003.

Yang, C.T., Sediment Transport: Theory and Practice, Int. Ed., Singapore, 1996.

Yang, C.T., 'Incipient motion and sediment transport', *Journal of Hydraulics Division*, ASCE, Vol. 99, No. HY10, Proceeding Paper 10067, pp. 1679–1704, 1973.

Yang, C.T., 'Unit stream power and sediment transport', *Journal of Hydraulics Division*, ASCE, Vol. 98, No. HY10, Proceeding Paper 9295, pp. 1805–1826, 1972.

Yılmaz, Ş., 'Prediction of Reservoir Siltation: Theory and Practice', 34th International Post-Graduate Course on Hydrology, Hungary, 2003.

Yücel, Ö. and Graf, W.H., 'Bed Load Deposition and Delta Formation: A mathematical Model', Fritz Engineering Laboratory Report 384.1, Lehigh University, Bethlehem, Pennsylvania, USA, 1973.

Zhang, R. and Xie, J., 'Sedimentation research in China, China Water and Power Press.' 1993.

Algorithm of delta and code list of extended delta program

Figure A.1 Decoupled simulation algorithm DELTA (Graf 2001)

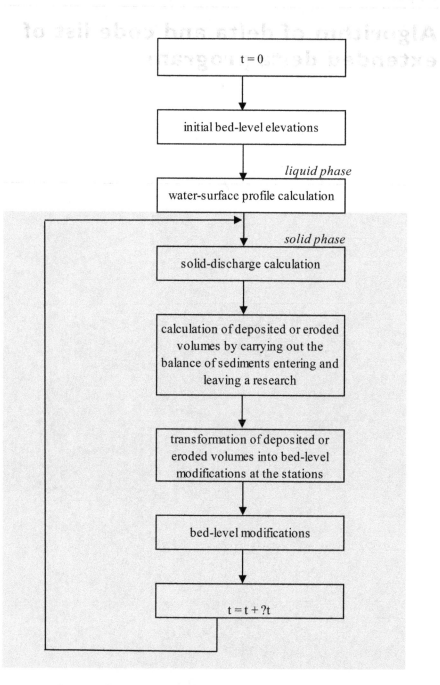

Figure A.1 Decoupled simulation algorithm DELTA (Graf, 2001)

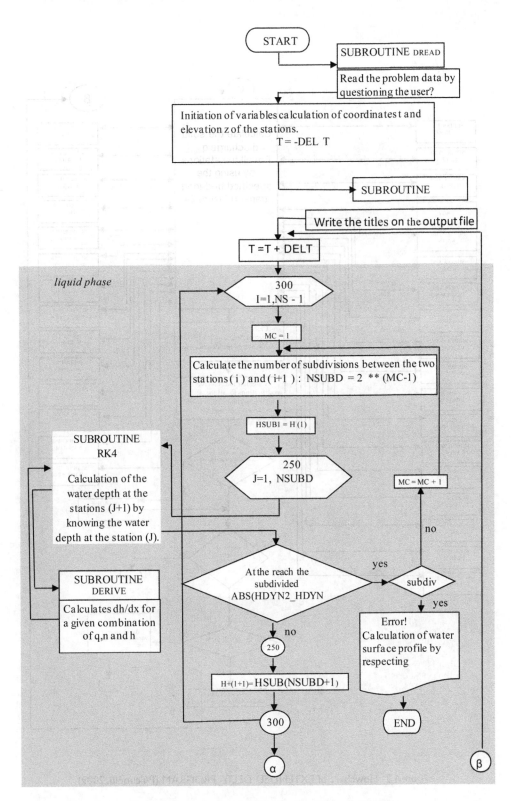

Figure A.2 Flowchart of DELTA (Graf and Altınakar, 1998)

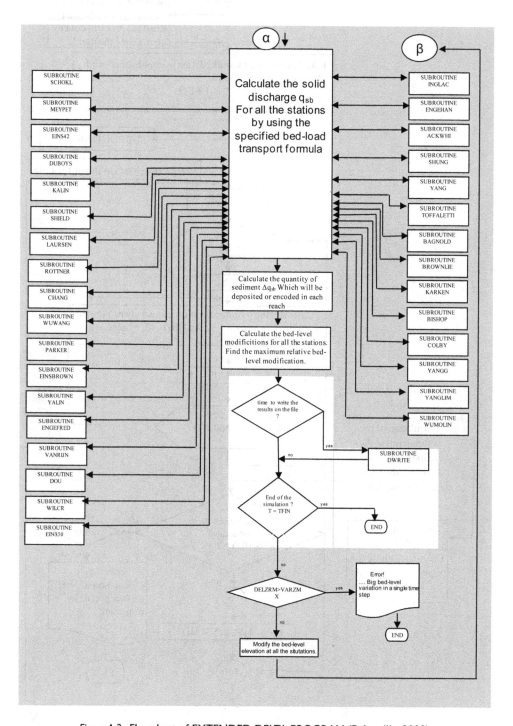

Figure A.3 Flowchart of EXTENDED DELTA PROGRAM (Pulcuoğlu, 2009)

```
PROGRAM DELTA
! THIS PROGRAM IS EXTENDED DELTA PROGRAM. F90COMPILER IS NEEDED.
! FOR THE DETAIL OF DELTA PROGRAM READER SHOULD SEE THE ORIGINAL !REFERENCES.
! HERE DELTA PROGRAM IS ADOPTED TO INLUDE DDITIONAL BED LOAD !EQUATION.
!
! MAIN PROGRAM FOR CALCULATING THE BED-LOAD TRANSPORT IN A
! RIVER-RESERVOIR SYSTEM BY TAKING INTO ACCOUNT THE MODIFICATION
! OF THE BED LEVEL.
!
! LIST OF VARIABLES DEFINED GLOBALLY FOR THE MAIN PROGRAM AND THE
! SUBPROGRAMS. VARIABLES DEFINED LOCALLY IN SUBPROGRAMS ARE LISTED AT
! THE BEGINNING OF EACH SUBPROGRAM.

! CFOI = COEFFICIENT OF VOLUME INCREASE DUE TO POROSITY
! CN = TOTAL MANNING COEFFICIENT
! CN50 = MANNING COEFFICIENT DUE TO GRAIN ROUGHNESS
! CPTT = TIME-STEP COUNTER
! CRAM = WEIGHTING COEFFICIENT FOR UPSTREAM STATION
!    USED IN CALCULATION OF DEPOSITION
! CRAV= WEIGHTING COEFFICIENT FOR DOWNSTREAM STATION
!    USED IN CALCULATION OF DEPOSITION
! D50= AVERAGE SEDIMENT DIAMETER
! DELQS (NSMAX-1) = SEDIMENTS DEPOSITED OR ERODED IN A REACH
!        BOUNDED BY TWO PRINCIPAL SECTIONS
! DELT = TIME STEP
! DELZ (NSMAX) = ARRAY CONTAINING THE BED LEVEL CHANGES AT ALL
!               STATIONS
! DELZMX = MAXIMUM BED-LEVEL CHANGE
! DELZRM = DIMENSIONLESS MAXIMUM BED-LEVEL CHANGE
! DHDYNM = MAXIMUM DIFFERENCE IN DYNAMIC HEAD (U2/2g)
!      BETWEEN TWO CONSECUTIVE STATIONS
! DQSDET = TOTAL VOLUME DEPOSITED SINCE T = 0
! DQSERT = TOTAL VOLUME ERODED SINCE T = 0
! DX = DISTANCE BETWEEN TWO PRINCIPAL STATIONS
! DXSUB = DISTANCE BETWEEN INTERPOLATED STATIONS
! FCOR = ROUGHNESS PARAMETER IN BED-LOAD FORMULA OF
!    MEYER-PETER (1948) (K/K'=n'/n)
! FICHS = NAME OF OUTPUT FILE
! FRNAM = FROUDE NUMBER AT UPSTREAM STATION OF A REACH
! FRNAV = FROUDE NUMBER AT DOWNSTREAM STATION OF A REACH
! G = GRAVITATIONAL ACCELERATION
! H(NSMAX) = WATER DEPTH AT PRINCIPAL STATIONS
! HAMHAV = RATIO OF UPSTREAM/DOWNSTREAM HEIGHTS OF TRAPE-
!    ZOID FORMED BY SEDIMENTS DEPOSITED IN A REACH
! HDYN1 = DYNAMIC HEAD (U^2/2G) AT STATION 1
! HDYN2 = DYNAMIC HEAD (U^2/2G) AT STATION 2
! HSUB (2**MCMAX+1) = WATER DEPTH AT INTERPOLATED STATIONS
! I = DO-LOOP COUNTER VARIABLE
! IAM = UPSTREAM STATION NUMBER FOR A GIVEN REACH
! IAV = DOWNSTREAM STATION NUMBER FOR A GIVEN REACH
! II = DO-LOOP COUNTER VARIABLE
! IMAXR = NUMBER OF THE STATION WHERE WE HAVE "DELZRM"
! IMAXZ = NUMBER OF THE STATION WHERE WE HAVE "DELZMX"
! J = DO-LOOP COUNTER VARIABLE
! MC = NUMBER OF SUBDIVISIONS (IN POWERS OF 2)
! MCMAX = MAXIMUM NUMBER OF SUBDIVISIONS (IN POWERS OF 2)
```

```
! ND = NUMBER OF PRINCIPAL REACHES
! NFTS = NUMBER OF BED-LOAD FORMULA TO BE USED
! NMC = NUMBER OF SUBDIVISIONS (IN POWERS OF 2)
!    SPECIFIED BY USER
! NOUT = UNIT NUMBER OF OUTPUT FILE
! NPP = PRINTING FREQUENCY
! NPT = PRINTING TIME COUNTER
! NS = NUMBER OF PRINCIPAL STATIONS
! NSMAX = NUMBER OF MAXIMUM STATIONS ALLOWED BY THE PROGRAM
! NSUBD = NUMBER OF SUBDIVISIONS AT A GIVEN REACH
! PI = PI NUMBER
! QCRIT = CRITICAL DISCHARGE IN FORMULA OF SCHOKLITSCH
! QSU(NSMAX) = SEDIMENTS TRANSPORTED THROUGH A STATION
! QU = UNIT DISCHARGE (WATER)
! QUIT = READ FOR TERMINATING THE PROGRAM
! ROE = DENSITY OF WATER
! ROS = DENSITY OF SEDIMENTS
! SEFF = SLOPE OF ENERGY GRADE LINE AT A STATION
! SF = INITIAL BED SLOPE
! SFTR = LOCAL BED SLOPE AT A GIVEN REACH
! SS = SPECIFIC DENSITY OF THE SEDIMENTS
! T = TIME
! TFIN = FINAL TIME
! TJ = TIME (NUMBER OF DAYS)
! TL = TOTAL LENGTH OF RIVER REACH STUDIED
! VARZMX = MAXIMUM RELATIVE BED-LEVEL CHANGE ALLOWED BY
!     THE PROGRAM IN A SINGLE TIME STEP
! X(NSMAX)= X-COORDINATES OF PRINCIPAL STATIONS
! X1 = X COORDINATE OF MOST DOWNSTREAM STATION
! XF = X COORDINATE OF MOST UPSTREAM STATION
! XSUB (2**MCMAX+1) = X-COORDINATES OF INTERPOLATED STATIONS
! ZF(NSMAX)= BED-LEVEL ELEVATIONS AT ANY TIME
! ZFI(NSMAX) = INITIAL BED-LEVEL ELEVATIONS
!
      PARAMETER (NOUT = 10 , NSMAX = 1000 , MCMAX = 7)
      COMMON /DONNE1/ SFTR, QU, CN
      EXTERNAL DERIVE
      CHARACTER*1 QUIT
      CHARACTER*40 FICHS
      DOUBLE PRECISION ZF (NSMAX), ZFI (NSMAX)
      DOUBLE PRECISION QSU(NSMAX), DELZ(NSMAX), DELQS(NSMAX-1)
      DOUBLE PRECISION DQSDET, DQSERT, DELZMX, DELZRM
      DOUBLE PRECISION DELT, TFIN, T
      DIMENSION X(NSMAX), H(NSMAX)
      DIMENSION XSUB(2**MCMAX+1), HSUB(2**MCMAX+1)
      OPEN (UNIT = 8, FILE = 'GRAPH.DAT', STATUS = 'NEW')
        CALL DREAD(NOUT, NSMAX, MCMAX,&
           QU, SF, CN, D50, ROS, ROE, SS, CN50,&
           X(1), XF, TL, DX, ND, NS, NMC,&
           H(1), DHDYNM, VARZMX,&
                                   DELT, TFIN ,&
           NFTS, CFOI, FCOR, HAMHAV, CRAM, CRAV,&
           FICHS, NPP )

      PI = 3.1415927
      G = 9.81
```

```
        T = -DELT
        CPTT = 0.0
        NPT = -1
        DQSDET = 0.0
        DQSERT = 0.0
        X(1) = 0.0
        ZF(1) = 0.0
        DO 10 I = 2, NS
        X(I) = X(I-1) + DX
        ZF(I) = ZF(I-1) + SF * DX
        ZFI(I) = ZF(I)
  10    CONTINUE
        CALL TITLES (NOUT, SF, D50, CN50, CN, ROS, ROE, QU,&
              DHDYNM, NMC, VARZMX, CFOI, HAMHAV, FCOR,&
              NFTS, X1, XF, TL, DX, ND, NS, H(1),&
              DELT, TFIN, NPP, FICHS)
 100    T = T + DELT
        CPTT = CPTT + 1.0
        NPT = NPT + 1
        TJ = T / 86400
        WRITE(*,110) CPTT, TJ
 110    FORMAT(/' TIME STEP = ',F15.0/&
              ' TIME (days) = ',F10.3)
        DO 300 I = 1, NS-1
        SFTR = (ZF(I+1) - ZF(I)) / DX
        MC = 0
 200    NSUBD = 2**MC
        DXSUB = (X(I+1) - X(I)) / NSUBD
        HSUB(1) = H(I)
        XSUB(1) = X(I)
        DO 250 J = 1, NSUBD
        XSUB(J+1) = X(I) + DXSUB * NSUBD
        CALL RK4(G, HSUB(J), DXSUB, HSUB(J+1), DERIVE)
        HDYN1 = QU**2 / (2*G*HSUB(J)**2)
        HDYN2 = QU**2 / (2*G*HSUB(J+1)**2)
        IF(ABS(HDYN2-HDYN1).GT.DHDYNM)THEN
           MC = MC+1
           WRITE(*,223) 2**MC, I, I+1
 223    FORMAT(/1X,I2,' subdivisions between sections ',I5,' and ',I5)
        IF(MC.GT.NMC)THEN
           FRNAM = QU / (H(I) * SQRT(G * H(I)))
           FRNAV = QU / (H(I+1) * SQRT(G * H(I+1)))
           WRITE(*,225) T , CPTT, I, I+1, DHDYNM, NMC,&
                FRNAM, FRNAV
        WRITE(NOUT,225) T, CPTT, I, I+1, DHDYNM, NMC,&
                FRNAM, FRNAV
 225    FORMAT(//&
        ' ERROR ! Computation of the backwater curve by respecting'/&
        ' the condition on Del(U^2/2g) is impossible.'/&
        ' Time (s) = ',D15.7/&
        ' Step no = ',F15.0/&
        ' Reach between sections = ',I5,' and',I5/&
                 ' Del(U^2/2g) tolerated (m) = ',F7.5/&
                 ' Number of calculated subdivisions =',I2/&
        ' Either increase the tolerated Del(U^2/2g) value,'/&
            ' or ask for more subdivisions'/&
```

```
          ' Verify also the Froude Number (Fr < 1):'/&
          ' Fr upstream section (-): ',F5.2/&
            ' Fr downstream section (-): ',F5.2//&
                   ' ABNORMAL END!'//)
          CALL DWRITE(NOUT, NS, T, X, ZF, ZFI, H, QU, G,&
             QSU, DELQS, DQSDET, DQSERT, CPTT,&
             DELZ, DELZMX, IMAXZ, DELZRM, IMAXR)
          WRITE(*,1000)
          READ(*,1001)QUIT
          STOP
        ENDIF
         GO TO 200
        ENDIF
250   CONTINUE
      H(I+1) = HSUB(NSUBD+1)
300   CONTINUE
       DO 400 II = 1, NS-1
       I = NS + 1 - II
       SEFF = (QU*CN)**2/H(I)**(10./3.)
       SFTR = (ZF(I) - ZF(I-1)) / DX
       IF(NFTS.EQ.1)THEN
! USE THE FORMULA OF SCHOKLITSCH (1950)
          CALL SCHOKL(QU, D50, SS, SEFF, QCRIT, QSU(I))
        ELSE IF(NFTS.EQ.2)THEN
! USE THE FORMULA OF MEYER-PETER et al. (1948)
          CALL MEYPET(G, SEFF, ROE, ROS, D50, FCOR, H(I), QSU(I))
        ELSE IF(NFTS.EQ.3)THEN
! USE THE FORMULA OF EINSTEIN (1942)
          CALL EINS42(SS, G, D50, H(I), SEFF, QSU(I))
        ELSE IF(NFTS.EQ.4)THEN
! USE THE FORMULA INGLAC PROGRAMMED BY THE USER
          CALL INGLAC(G,D50,SS,QU,H(I),QSU(I))
        ELSE IF(NFTS.EQ.5)THEN
! USE THE FORMULA ENGEHAN PROGRAMMED BY THE USER
          CALL ENGEHAN (ROE,G,H(I),SEFF,QU,SF,SS,D50,QSU(I))
        ELSE IF(NFTS.EQ.6)THEN
!USE THE FORMULA DUBOY PROGRAMMED BY THE USER
          CALL DUBOY(ROE,G,H(I),SEFF,D50,SS,QSU(I))
        ELSE IF(NFTS.EQ.7)THEN
! USE THE FORMULA KALIN PROGRAMMED BY THE USER
          CALL KALIN(ROE,G,H(I),SEFF,D50,SS,QSU(I))
        ELSE IF(NFTS.EQ.8)THEN
! USE THE FORMULA SHIELD PROGRAMMED BY THE USER
          CALL SHIELD (G,ROE,H(I),SEFF,D50,SS,QU,SF,QSU(I))
        ELSE IF(NFTS.EQ.9)THEN
! USE THE FORMULA LAURSEN PROGRAMMED BY THE USER
          CALL LAURSEN(G,D50,SS,QU,H(I),ROE,QSU(I))
        ELSE IF(NFTS.EQ.10)THEN
! USE THE FORMULA ACKWHI PROGRAMMED BY THE USER
          CALL ACKWHI (QU,H(I),G,D50,SS,ROE,SEFF,QSU(I))
        ELSE IF(NFTS.EQ.11)THEN
! USE THE FORMULA SHUNG PROGRAMMED BY THE USER
          CALL SHUNG(G,D50,SS,QU,H(I),SF,QSU(I))
        ELSE IF(NFTS.EQ.12)THEN
! USE THE FORMULA YANG PROGRAMMED BY THE USER
          CALL YANG(G,D50,SS,QU,H(I),ROE,SEFF,QSU(I))
```

```
      ELSE IF(NFTS.EQ.13)THEN
! USE THE FORMULA ROTTNER PROGRAMMED BY THE USER
      CALL ROTTNER (QU,H(I),D50,SS,G,QSU(I))
      ELSE IF(NFTS.EQ.14)THEN
! USE THE FORMULA TOFFALETI PROGRAMMED BY THE USER
      CALL TOFFALETI (G,D50,SS,QU,H(I),SF,QSU(I))
      ELSE IF(NFTS.EQ.15)THEN
! USE THE FORMULA CHANG PROGRAMMED BY THE USER
      CALL CHANG(QU,H(I),ROE,G,SEFF,D50,SS,SF,QSU(I))
      ELSE IF(NFTS.EQ.16)THEN
! USE THE FORMULA BAGNOLD PROGRAMMED BY THE USER
      CALL BAGNOLD(QU,H(I),ROE,G,SEFF,D50,SS,QSU(I))
      ELSE IF(NFTS.EQ.17)THEN
! USE THE FORMULA WUWANG PROGRAMMED BY THE USER
      CALL WUWANG(G,D50,SS,QU,H(I),QSU(I))
      ELSE IF(NFTS.EQ.18)THEN
! USE THE FORMULA BROWNLIE PROGRAMMED BY THE USER
      CALL BROWNLIE(QU,H(I),G,D50,SS,SF,QSU(I))
      ELSE IF(NFTS.EQ.19)THEN
! USE THE FORMULA PARKER PROGRAMMED BY THE USER
      CALL PARKER(ROE,G,H(I),SEFF,SS,D50,QSU(I))
      ELSE IF(NFTS.EQ.20)THEN
! USE THE FORMULA EINSBROWN PROGRAMMED BY THE USER
      CALL EINSBROWN(G,D50,SS,QU,H(I),ROE,SEFF,QSU(I))
      ELSE IF(NFTS.EQ.21)THEN
! USE THE FORMULA YALINPROGRAMMED BY THE USER
      CALL YALIN(ROE,G,H(I),SEFF,D50,SS,QSU(I))
      ELSE IF(NFTS.EQ.22)THEN
! USE THE FORMULA ENGEFRED PROGRAMMED BY THE USER
      CALL ENGEFRED(ROE,G,H(I),SEFF,D50,SS,QSU(I))
      ELSE IF(NFTS.EQ.23)THEN
! USE THE FORMULA VANRIJN PROGRAMMED BY THE USER
      CALL VANRIJN(QU,H(I),D50,SS,G,ROE,QSU(I))
      ELSE IF(NFTS.EQ.24)THEN
! USE THE FORMULA DOU PROGRAMMED BY THE USER
      CALL DOU(H(I),D50,SS,G,QU,SF,QSU(I))
      ELSE IF(NFTS.EQ.25)THEN
! USE THE FORMULA KARKEN PROGRAMMED BY THE USER
      CALL KARKEN(SEFF,SS,G,D50,QU,H(I),ROE,QSU(I))
      ELSE IF(NFTS.EQ.26)THEN
! USE THE FORMULA BISHOP PROGRAMMED BY THE USER
      CALL BISHOP(SEFF,SS,D50,H(I),G,QSU(I))
      ELSE IF(NFTS.EQ.27)THEN
! USE THE FORMULA WILCR PROGRAMMED BY THE USER
      CALL WILCR(G,ROE,H(I),SEFF,SS,D50,QSU(I))
      ELSE IF(NFTS.EQ.28)THEN
! USE THE FORMULA EIN50 PROGRAMMED BY THE USER
      CALL EINS50(G,H(I),SEFF,D50,ROS,ROE,SS,QSU(I))
      ELSE IF(NFTS.EQ.29)THEN
! USE THE FORMULA COLBY PROGRAMMED BY THE USER
      CALL COLBY(QU,H(I),D50,QSU(I),G,ROS)
      ELSE IF(NFTS.EQ.30)THEN
! USE THE FORMULA YANGG PROGRAMMED BY THE USER
      CALL YANGG(G,D50,SS,QU,H(I),ROE,SEFF,QSU(I))
      ELSE IF(NFTS.EQ.31)THEN
! USE THE FORMULA YANGLIM PROGRAMMED BY THE USER
```

```
       CALL  YANGLIM(ROE,G,H(I),SEFF,D50,SS,QSU(I),ROS)
       ELSE IF(NFTS.EQ.32)THEN
! USE THE FORMULA WUMOLIN PROGRAMMED BY THE USER
       CALL  WUMOLIN(QU,H(I),G,D50,SS,QSU(I))
       ELSE
       WRITE(*,450)NFTS
       WRITE(NOUT,450)NFTS
450    FORMAT(' ERROR ! TRANSPORT FORMULA NFTS = ',I2,'DOES NOT'/&
       ' EXIST. VERIFY THE INPUT DATA'//&
       ' ABNORMAL END'//)
       WRITE(*,1000)
       READ(*,1001)QUIT
       STOP
       ENDIF
400    CONTINUE
       QSU(1) = 0.0
       DO 500 II = 1, NS-1
       IAM = NS + 1 - II
       IAV = IAM - 1
       DELQS(IAV) = QSU(IAM) - QSU(IAV)
       IF(DELQS(IAV).GE.0.0)THEN
        DQSDET = DQSDET + DELQS(IAV) * DELT
       ELSE
        DQSERT = DQSERT + DELQS(IAV) * DELT
       ENDIF
500    CONTINUE
       DELZ(1) = DELQS(1) * DELT * CFOI * CRAV / DX
       DELZMX = DELZ(1)
       DELZRM = DELZ(1) / H(1)
       IMAXZ = 1
       IMAXR = 1
       DO 600 I = 2, NS-1
       DELZ(I) = (DELQS(I) * CRAV + DELQS(I-1) * CRAM) * DELT / (DX * 2.)
       DELZ(I) = DELZ(I) * CFOI
       IF(DELZMX.LT.DELZ(I))THEN
        DELZMX = DELZ(I)
        IMAXZ = I
       ENDIF
       IF(DELZRM.LT.DELZ(I)/H(I))THEN
        DELZRM = DELZ(I) / H(I)
        IMAXR = I
       ENDIF
600    CONTINUE
       DELZ(NS) = DELQS(NS-1) * DELT * CFOI / DX
       IF(DELZMX.LT.DELZ(NS))THEN
        DELZMX = DELZ(NS)
        IMAXZ = NS
       ENDIF
       IF(DELZRM.LT.DELZ(NS)/H(NS))THEN
        DELZRM = DELZ(NS) / H(NS)
        IMAXR = NS
       ENDIF
       IF(CPTT.EQ.1.0.OR.NPT.EQ.NPP.OR.T.GE.TFIN)THEN
          CALL DWRITE(NOUT, NS, T, X, ZF, ZFI, H, QU, G,&
          QSU, DELQS, DQSDET, DQSERT, CPTT,&
          DELZ, DELZMX, IMAXZ, DELZRM, IMAXR)
```

```
      IF(NPT.EQ.NPP)NPT = 0
      ENDIF
      IF(T.GE.TFIN)GO TO 999
      IF(DELZRM.GT.VARZMX)THEN
        WRITE(*,650) T, DELZMX, DELZRM, IMAXR
        WRITE(NOUT,650) T, DELZMX, DELZRM, IMAXR
650   FORMAT(' ERROR ! Bed-level variation at a single step is too'/&
        ' large. Use a smaller time step'/&
        ' T (s) = ',D15.7/&
        ' DELZMX (m) = ',D12.6/&
        ' DELZRM (-) = ',F10.6/&
        ' IMAX = ',I6//&
        ' ABNORMAL END !'//)
        CALL DWRITE(NOUT, NS, T, X, ZF, ZFI, H, QU, G,&
              QSU, DELQS, DQSDET, DQSERT, CPTT,&
              DELZ, DELZMX, IMAXZ, DELZRM, IMAXR)
        WRITE(*,1000)
        READ(*,1001)QUIT
        STOP
      ENDIF
      DO 700 I = 1,NS
      ZF(I) = ZF(I) + DELZ(I)
700   CONTINUE
      GO TO 100

999 WRITE(*,1000)
1000 FORMAT(/' NORMAL END OF PROGRAM.'//&
        ' PRESS RETURN TO END THE PROGRAM.'/)
      READ(*,1001)QUIT
1001 FORMAT(A)
      STOP
      END

      SUBROUTINE DREAD(NOUT, NSMAX, MCMAX,&
            QU, SF, CN, D50, ROS, ROE, SS, CN50,&
            X1, XF, TL, DX, ND, NS, NMC,&
            H1, DHDYNM, VARZMX,&
            DELT, TFIN,&
            NFTS, CFOI, FCOR, HAMHAV, CRAM, CRAV,&
            FICHS, NPP)
!
! ANSWER = ANSWER (Yes OR No) TO A QUESTION
! H1 = WATER DEPTH AT THE MOST DOWNSTREAM STATION
!      (CONTROL SECTION)
! POROS = POROSITY OF DEPOSITED SEDIMENTS
! DECLARATION OF VARIABLES
      CHARACTER*1 ANSWER
      CHARACTER*40 FICHS
      DOUBLE PRECISION DELT, TFIN
      OPEN(UNIT = 7, FILE = 'DIALOG.DAT', STATUS = 'NEW')
      WRITE(*,10)
      WRITE(7,10)
10    FORMAT(//' PROGRAM FOR BED-LOAD SEDIMENT TRANSPORT CALC&
              ULATION'/&
            ' BY TAKING INTO ACCOUNT BED-LEVEL MODIFICAT&
                IONS.'//&
```

```
           ' NOTES :'/' - UNIT SYSTEM = SI'/&
           ' - NUM. METHOD FOR WATER SURFACE PROFILE CALCULATION = 4&
                th ORDER RUNGE-KUTTA'/&
           ' - THE FLOW BEING SUBCRITICAL (Fr < 1), THE WATER-SURFA&
                CE PROFILE CALCULATION'/&
           ' STARTS AT THE DOWNSTREAM END AND PROGRESSES TOWARDS U&
                PSTREAM'/&
           ' - SEDIMENT TRANSPORT CALCULATIONS ARE CARRIED OUT FROM &
                CUPSTREAM TO DOWNSTREAM'/&
           ' - THE CALCULATIONS ARE MADE FOR A UNIT WIDTH'//)
        WRITE(*,90)
        WRITE(7,90)
90   FORMAT(' PHYSICAL CHARACTERISTICS DATA :')
100  WRITE(*,110)
110  FORMAT(' Initial bed slope, SF (-) ? = ',$)
        READ(*,*,ERR=100)SF
        WRITE(7,110)
        WRITE(7,*)SF
!
120  WRITE(*,130)
130  FORMAT(/' Average diameter of sediments, D50 (mm) ? = ',$)
        READ(*,*,ERR=120)D50
        WRITE(7,130)
        WRITE(7,*)D50
        D50 = D50 * 0.001
!
        CN50 = D50**(1./6.) / 21.1
140  WRITE(*,150)CN50
150  FORMAT(//' According to eq. 3.18, the Manning coefficient'/&
           ' due to the grain roughness is :'//&
           ' CN50 = d50^(1/6) / 21.1 = ',F6.4,'(s/m^1/3)'//&
           ' Total Manning-Strickler coeff., CN (s/m^1/3) ? = ',$)
        READ(*,*,ERR=140)CN
        IF(CN.LT.CN50)THEN
        WRITE(*,155)
        WRITE(7,155)
155      FORMAT(//' ERROR ! CN should be greater than CN50'/)
        GO TO 140
        ENDIF
        WRITE(7,150)
        WRITE(7,*)CN50
160  WRITE(*,170)
170  FORMAT(/' Density of sediments, ROS (kg/m3) ? = ',$)
        READ(*,*,ERR=160)ROS
        WRITE(7,170)
        WRITE(7,*)ROS
180  WRITE(*,190)
190  FORMAT(' Density of water, ROE (kg/m3) ? = ',$)
        READ(*,*,ERR=180)ROE
        WRITE(7,190)
        WRITE(7,*)ROE
        SS = ROS / ROE
200  WRITE(*,210)
210  FORMAT(/' Unit discharge, QU (m2/s) ? = ',$)
        READ(*,*,ERR=200)QU
        WRITE(7,210)
```

```
      WRITE(7,*)QU
300   WRITE(*,310)

310   FORMAT(//' CHOICE OF BED-LOAD TRANSPORT FORMULA :'/&
      ' The program allows the use of one of the'/&
      ' following 27 bed-load formulae:'/&
      ' 1- Schoklitsch (1950)'/&
      ' 2- Meyer-Peter et al. (1948)'/&
      ' 3- Einstein (1942)'/&
          ' 4- Inglis-Lacey Formula (Inglis-1968)'/&
      ' 5- Engelund-Hansen Approach'/&
      ' 6- DuBoys Formula'/&
      ' 7- Kalinske (1947)'/&
      ' 8- Shields Bed-load Equation'/&
      ' 9- Laursen Approach '/&
      '10-Ackers and White Formula'/&
      '11-Shen and Hung Method'/ &
      '12-Yangs Sand Formula (1973)'/&
      '13-Rottners Formula'/&
      '14-Toffaleti Procedure (1969-Total Load)'/&
      '15-Chang-Simon-Richardson Approach'/&
      '16-Bagnolds Equation'/&
       '17-Wu-Wang-Jia Equation (2000)'/&
      '18-Brownlies Method (1981)'/&
      '19-Parkers Approach (1990)'/&
      '20-Einstein-Brown Approach (1950)'/&
      '21-Yalins Formula (1972)'/&
      '22-Engelund-Fredsoe (1976) '/&
      '23-Van Rijns Formula (1984)'/&
      '24-Dou Equation (1977)'/&
      '25-Karim-Kennedy Approach (1990)'/&
      '26-Bishop et al. Method (1965)'/&
      '27-Wilcock and Crowe Method (2003)'/&
      '28-Einstein Method (1950-Bed-load only)'/ &
      '29-Colby Relation (1964)'/&
      '30-Yangs Gravel Formula (1984)'/&
      '31-Yang-Lim Total-Load Formula (2003)'/&
      '32-Wu and Molinas Equation (2001)'/&
      ' Which formula do you choose ? (1 a 32) = ',$)
      READ(*,*,ERR=300)NFTS
      WRITE(7,310)
      WRITE(7,*)NFTS
      IF(NFTS.EQ.1)THEN
320   WRITE(*,330)
      WRITE(7,330)
330   FORMAT(/' SCHOKLITSCH (1950) bed-load formula is chosen.'/)
      ELSE IF(NFTS.EQ.2)THEN
      WRITE(*,340)
      WRITE(7,340)
340   FORMAT(/' MEYER-PETER (1948) bed-load formula is chosen.')
      IF(CN.GT.CN50)THEN
      WRITE(*,342)
342   FORMAT(/' Do you want to use the roughness parameter, FCOR?'/&
          ' Answer Y(es) or N(o)/CR ? = ',$)
      READ(*,345)ANSWER
345   FORMAT(A)
```

```
      WRITE(7,342)
      WRITE(7,345)ANSWER
      IF(ANSWER.EQ.'Y'.OR.ANSWER.EQ.'y')THEN
       FCOR = (CN50 / CN)**1.5
      ELSE
       FCOR = 1.0
      ENDIF
      ELSE
       FCOR = 1.0
      ENDIF
      WRITE(*,346) FCOR
346   FORMAT(/' The roughness parameter is then, FCOR = (CN50 / CN)*&
                    1.5 = ',F6.3/)
      WRITE(7,346) FCOR
          ELSE IF(NFTS.EQ.3)THEN
      WRITE(*,350)
      WRITE(7,350)
350   FORMAT(/' EINSTEIN (1942) bed-load formula is chosen.'/)
          ELSE IF(NFTS.EQ.4)THEN
      WRITE(*,360)
      WRITE(7,360)
360   FORMAT(/' Inglis-Lacey (Inglis 1968) Formula is chosen.'/)

      ELSE IF(NFTS.EQ.5)THEN
       WRITE(*,361)
       WRITE(7,361)
361   FORMAT(/' Engelund-Hansen Approach is chosen.'/)

      ELSE IF(NFTS.EQ.6)THEN
        WRITE(*,362)
       WRITE(7,362)
362   FORMAT(/' Duboys Formula is chosen.'/)

      ELSE IF(NFTS.EQ.7)THEN
       WRITE(*,363)
       WRITE(7,363)
363   FORMAT(/' Kalinske (1947) Formula is chosen.'/)

      ELSE IF(NFTS.EQ.8)THEN
        WRITE(*,364)
       WRITE(7,364)
364   FORMAT(/' Shields Bedload Equation is chosen.'/)

      ELSE IF(NFTS.EQ.9)THEN
        WRITE(*,365)
       WRITE(7,365)
365   FORMAT(/' Laursen Equation is chosen.'/)

      ELSE IF(NFTS.EQ.10)THEN
        WRITE(*,366)
       WRITE(7,366)
366   FORMAT(/' Ackers and White Formula is chosen.'/)

      ELSE IF(NFTS.EQ.11)THEN
        WRITE(*,367)
       WRITE(7,367)
```

```
367  FORMAT(/' Shen and Hung Method is chosen.'/)


     ELSE IF(NFTS.EQ.12)THEN
        WRITE(*,368)
        WRITE(7,368)
368  FORMAT(/' Yangs Sand Formula is chosen.'/)


     ELSE IF(NFTS.EQ.13)THEN
        WRITE(*,369)
        WRITE(7,369)
369  FORMAT(/' Rottners Formula is chosen.'/)


     ELSE IF(NFTS.EQ.14)THEN
        WRITE(*,370)
        WRITE(7,370)
370  FORMAT(/' Toffaleti Procedure (1969-Bed load Part) is chosen.'/)


     ELSE IF(NFTS.EQ.15)THEN
        WRITE(*,371)
        WRITE(7,371)
371  FORMAT(/' Chang-Simon-Richardson Approach is chosen.'/)


     ELSE IF(NFTS.EQ.16)THEN
        WRITE(*,372)
        WRITE(7,372)
372  FORMAT(/' Bagnolds Equation is chosen.'/)


     ELSE IF(NFTS.EQ.17)THEN
        WRITE(*,373)
        WRITE(7,373)
373  FORMAT(/' Wu-Molinas Equation (2000) is chosen.'/)


     ELSE IF(NFTS.EQ.18)THEN
        WRITE(*,374)
        WRITE(7,374)
374  FORMAT(/' Brownlies Method (1981) is chosen.'/)


     ELSE IF(NFTS.EQ.19)THEN
        WRITE(*,375)
        WRITE(7,375)
375  FORMAT(/' Parkers Approach (1990) is chosen.'/)


     ELSE IF(NFTS.EQ.20)THEN
        WRITE(*,376)
        WRITE(7,376)
376  FORMAT(/' Einstein-Brown Approach (1950) is chosen.'/)


     ELSE IF(NFTS.EQ.21)THEN
        WRITE(*,377)
        WRITE(7,377)
377  FORMAT(/' Yalins Formula (1972) is chosen.'/)


     ELSE IF(NFTS.EQ.22)THEN
        WRITE(*,378)
        WRITE(7,378)
378  FORMAT(/' Engelund-Fredsoe (1976) is chosen.'/)
```

```
      ELSE IF(NFTS.EQ.23)THEN
        WRITE(*,379)
        WRITE(7,379)
379   FORMAT(/' Van Rijns Formula (1984) is chosen.'/)

      ELSE IF(NFTS.EQ.24)THEN
        WRITE(*,555)
        WRITE(7,555)
555   FORMAT(/' Dou Equation (1977) is chosen.'/)

      ELSE IF(NFTS.EQ.25)THEN
        WRITE(*,556)
        WRITE(7,556)
556   FORMAT(/' Karim-Kennedy (1990) Approach is chosen.'/)

      ELSE IF(NFTS.EQ.26)THEN
        WRITE(*,557)
        WRITE(7,557)
557   FORMAT(/' Bishop et al. Method (1965) is chosen.'/)

      ELSE IF(NFTS.EQ.27)THEN
        WRITE(*,558)
        WRITE(7,558)
558   FORMAT(/' Wilcock and Crowe Method (2003) is chosen.'/)

      ELSE IF(NFTS.EQ.28)THEN
        WRITE(*,559)
        WRITE(7,559)
559   FORMAT(/' Einstein Method (1950) (bed-load part only) is chosen&
        .'/)

      ELSE IF(NFTS.EQ.29)THEN
        WRITE(*,560)
        WRITE(7,560)
560   FORMAT(/' Colby Relation (1964) is chosen.'/)

      ELSE IF(NFTS.EQ.30)THEN
        WRITE(*,561)
        WRITE(7,561)
561   FORMAT(/' Yangs Gravel Formula (1984) is chosen.'/)
      ELSE IF(NFTS.EQ.31)THEN
        WRITE(*,562)
        WRITE(7,562)
562   FORMAT(/' Yang-Lim Total-Load Formula (2003) is chosen.'/)
      ELSE IF(NFTS.EQ.32)THEN
        WRITE(*,563)
        WRITE(7,563)
563   FORMAT(/' Wu-Molinas Equation (2001) is chosen.'/)
      ELSE
        WRITE(*,380)
        WRITE(7,380)
380   FORMAT(' ERROR !... NON EXISTING CHOICE !'//)
      GO TO 300
    ENDIF
385   WRITE(*,390)
```

```
390   FORMAT(/' BED-LEVEL MODIFICATION DATA :'/&
          ' The bed-level variation during a single time step'/&
          ' may not be too big; otherwise, instabilities of'/&
          ' the bed profile may appear. At the end of each time'/&
          ' step, the program calculates the maximum relative bed-'/&
          ' level variation, DELZ/H, (bed-level variation divided'/&
          ' by the water depth) and checks that it is not bigger'/&
          ' than a value specified by the user.'//&
          ' Max. rel. bed-level variation, VARZMX (-) ? = ',$)
      READ(*,*,ERR=385)VARZMX
      WRITE(7,390)
      WRITE(7,*)VARZMX
391   WRITE(*,392)
392   FORMAT(/' Porosity of deposited sediments, POROS (-) ? = ',$)
      READ(*,*,ERR=391)POROS
      WRITE(7,392)
      WRITE(7,*)POROS
      CFOI = 1.0 / (1.0 - POROS)
!
395   WRITE(*,396)
396   FORMAT(/' The volume of sediments deposited in a reach is'/&
          ' transformed into a bed-level variation height at the'/&
          ' downstream and upstream, by defining a trapezoid.'/&
          ' The user controls the sediment distribution by choosing'/&
          ' the ratio of upstream/downstream heights of the trapezoid.'/&
          ' We recommend to use the values: 0.5 < HAMHAV < 1.'//&
          ' Ratio of upstr./downstr. heights, HAMHAV (-) ? = ',$)
      READ(*,*,ERR=395)HAMHAV
      WRITE(7,396)
      WRITE(7,*)HAMHAV

      CRAM = 2.0 * HAMHAV / (1 + HAMHAV)
      CRAV = 2.0 / (1 + HAMHAV)
      WRITE(*,400)
      WRITE(7,400)
400   FORMAT(//' INFORMATION ON COMPUTATIONAL DOMAIN :')
405   WRITE(*,410)
410   FORMAT(' x-coordinate of first station, X1 (m) ? = ',$)
      READ(*,*,ERR=405)X1
      WRITE(7,410)
      WRITE(7,*)X1
420   WRITE(*,430)
430   FORMAT(' x-coordinate of last station, XF (m) ? = ',$)
      READ(*,*,ERR=420)XF
      WRITE(7,430)
      WRITE(7,*)XF
      TL = XF - X1
440   WRITE(*,450)TL
450   FORMAT(/' Total reach length is therefore, TL (m) = ',&
      F10.2//&
          ' Now you have to specify the step length in x-direction.'/&
          ' If the step length is too long to guarantee a correct'/&
          ' prediction of the water-surface profile, the program'/&
          ' will automatically add some intermediate stations. The'/&
          ' results, however, are only printed at the stations'/&
          ' specified by the user; others remain invisible. The step'/&
```

```
          ' length for the space must therefore be specified to'/&
          ' guarantee a correct representation of physical processes'/&
          ' involved in the bed-load transport. In case of doubt, you'/&
          ' can repeat the simulation with different step lengths and'/&
          ' compare the results.'//&
          ' Step length in x-direction, DX (m) ? = ',$)
      READ(*,*,ERR=440)DX
      WRITE(7,450)TL
      WRITE(7,*)DX
      ND = (XF -X1) / DX
      NS = ND + 1
      IF(NS.GT.NSMAX)THEN
       WRITE(*,460) NS, NSMAX
       WRITE(7,460) NS, NSMAX
460   FORMAT(//' ERROR ! The maximum number of principal sections'/&
          ' is defined as: NSMAX = ',I5/ &
          ' If you want to work with more principal'/&
          ' sections, you should change the parameter'/&
          ' NSMAX and recompile the program.')
       GO TO 440
      ENDIF
      WRITE(*,463) ND, NS
      WRITE(7,463) ND, NS
463   FORMAT(/' Number of reaches, ND = ',I5/&
          ' Number of stations, NS = ',I5/)
465   WRITE(*,470)
470   FORMAT(' Max. tolerated var. of dyn. head, DHDYNM (m) ? = ',$)
      READ(*,*,ERR=465)DHDYNM
      WRITE(7,470)
      WRITE(7,*)DHDYNM
475   WRITE(*,480) MCMAX, 2**MCMAX
      WRITE(7,480) MCMAX, 2**MCMAX
480   FORMAT(/' In case this value is exceeded, the reach will be'/&
          ' subdivided in order to refine the calculations.'/&
          ' The number of divisions is specified in powers of 2.'/&
          ' The maximum value is MCMAX = ',I3/&
          ' Which corresponds to 2.0^MCMAX = ',I3,' subdivisions.')
      WRITE(*,485)
485   FORMAT(/' Maximum number of subdiv. in powers of 2, NMC ? = ',$)
      READ(*,*,ERR=475)NMC
      IF(NMC.GT.MCMAX)GO TO 475
      WRITE(7,485)
      WRITE(7,*)NMC
     WRITE(*,490)
     WRITE(7,490)
490   FORMAT(//' INFORMATION ON BOUNDARY CONDITIONS :')
500   WRITE(*,510)
510   FORMAT(' Note that the sediment transport at downstream end'/&
          ' is automatically taken as zero, QSU(1) = 0.0'//&
          ' Water depth at downstream end, H1 (m) ? = ',$)
      READ(*,*,ERR=500)H1
      WRITE(7,510)
      WRITE(7,*)H1
! READ THE TIME STEP AND THE COMPUTATION TIME
      WRITE(*,590)
      WRITE(7,590)
```

```
590   FORMAT(//' PARAMETERS RELATED TO TIME AND PRINTING OF RESULTS :')
600   WRITE(*,610)
610   FORMAT(' Time step, DELT (days) ? = ',$)
      READ(*,*,ERR=600)DELT
      WRITE(7,610)
      WRITE(7,*)DELT
      DELT = DELT * 24.0 * 60.0 * 60.0
620   WRITE(*,630)
630   FORMAT(/' Duration of the simulation, TFIN (days) ? = ',$)
      READ(*,*,ERR=620)TFIN
      WRITE(7,630)
      WRITE(7,*)TFIN
      TFIN = TFIN * 24.0 * 60.0 * 60.0
850   WRITE(*,860)
860   FORMAT(/' Results will be printed every NPP step ? = ',$)
      READ(*,*,ERR=850)NPP
      WRITE(7,860)
      WRITE(7,*)NPP
900   WRITE(*,910)
910   FORMAT(/' Name of output file (max. 40 char.) ? = ',$)
      READ(*,345)FICHS
      OPEN( UNIT = NOUT, FILE = FICHS, STATUS = 'NEW', ERR = 900 )
      WRITE(7,910)
      WRITE(7,345)FICHS
      RETURN
      END
      SUBROUTINE RK4(G, Y, DX, YP, DERIVE)

! CALCULATE THE VALUES OF DEPENDENT VARIABLES AT STATION X+DX
! LIST OF VARIABLES LOCALLY DEFINED FOR SUBROUTINE SUBPROGRAM "RK4"
! TYPE NAME DIMEN. EXPLANATIONS
! DX2 = DX*0.5
! DX6 = DX/6
! K1 = K1 IN THE 4TH ORDER RUNGE-KUTTA METHOD
! K2 = K2 IN THE 4TH ORDER RUNGE-KUTTA METHOD
! K3 = K3 IN THE 4TH ORDER RUNGE-KUTTA METHOD
! K4 = K4 IN THE 4TH ORDER RUNGE-KUTTA METHOD
! Y = WATER DEPTH KNOWN AT THE DOWNSTREAM STATION
!     OF A REACH
! YP = WATER DEPTH CALCULATED AT THE UPSTREAM STATION
!     OF A REACH
! YT = TEMPORARY VALUE OF Y AT (X+DX/2) AND/OR AT
!     (X+DX) USED FOR CALCULATING K2, K3 AND K4
! DECLARATION OF VARIABLES
      REAL K1, K2, K3, K4
      DX2=DX*0.5
      DX6=DX/6.0
      CALL DERIVE( G, Y, K1)
      YT = Y + DX2 * K1
      CALL DERIVE(G, YT, K2)
      YT = Y + DX2 * K2
      CALL DERIVE(G, YT, K3)
      YT = Y + DX * K3
      CALL DERIVE(G, YT, K4)
      YP = Y + DX6 * (K1 + 2*K2 + 2*K3 + K4)
      RETURN
```

```
      END

      SUBROUTINE DERIVE( G, Y, DYDX)

! DYDX = THE DERIVATIVE dh/dx CALCULATED ACCORDING TO
!     DIFFERENTIAL EQUATION OF GRADUALLY VARIED FLOW
! Y = WATER DEPTH AT THE STATION WHERE dh/dx IS BEING
!     CALCULATED (REPLACES ALTERNATELY Y AND YT)
!
      COMMON /DONNE1/ SFTR, QU, CN
      DYDX = SFTR - (QU*CN)**2 / Y**(10./3.)
      DYDX = -DYDX / (1 - QU**2 / (G * Y**3))
!
      RETURN
      END

      SUBROUTINE TITLES( NOUT, SF, D50, CN50, CN, ROS, ROE, QU,&
              DHDYNM, NMC, VARZMX, CFOI, HAMHAV , FCOR,&
              NFTS, X1, XF, TL, DX, ND, NS, H1,&
              DELT, TFIN, NPP, FICHS)
      CHARACTER*40 FICHS
      CHARACTER*131 TITLE(25)
      DOUBLE PRECISION DELT, TFIN
!
! DATA
      DATA TITLE(1) /' PHYSICAL CHARACTERISTICS DATA :&
              CHOICE OF BED-LOAD FORMULA :'/
      DATA TITLE(2) /' Initial bed slope, SF (-) &
      = '/
      DATA TITLE(3) /' Average diameter of sediments, D50 (mm) &
      = '/
      DATA TITLE(4) /' Manning coeff. for sed. grains, CN50 (s/m^1/3) &
      = '/
      DATA TITLE(5) /' Manning-Strickler coefficient, CN (s/m1/3) &
      = '/
      DATA TITLE(6) /' Density of sediments, ROS (kg/m3) &
      = '/
      DATA TITLE(7) /' Density of water, ROE (kg/m3) &
      = '/
      DATA TITLE(8) /' Unit discharge, QU (m2/s) &
      = '/
      DATA TITLE(9) /' '/
      DATA TITLE(10)/' INFORMATION RELATED TO WATER-SURFACE CALCULATIONS &
              : INFORMATION RELATED TO THE CALCULATION OF DEPOSIT &
              ION VOLUME : '/
      DATA TITLE(11)/' Max. tolerated var. of dyn. head, DHDYNM (m) &
              = Max. rel. bed-level variation, VARZMX (-) &
              = '/
      DATA TITLE(12)/' Maximum number of subdiv. in powers of 2, NMC &
              = Coefficient of swelling, CFOI (-) &
              = '/
      DATA TITLE(13)/' &
              Ratio of upst./dwnst. heights, HAMHAV (-) &
              '/
      DATA TITLE(14)/' '/
      DATA TITLE(15)/' INFORMATION ON CALCULATION DOMAIN : &
```

```
              BOUNDARY CONDITIONS: '/
      DATA TITLE(16)/' x-coordinate of first station, X1 (m) &
                = The bed-load transport at the downstream end is zero. '/
      DATA TITLE(17)/' x-coordinate of last station, XF (m) &
                = Water depth at the downstream end, H1 (m) &
                = '/
      DATA TITLE(18)/' Total reach length is therefore, TL (m) &
                = '/
      DATA TITLE(19)/' Step length in x-direction, DX (m) &
                = PARAMETERS RELATED TO TIME: '/
      DATA TITLE(20)/' Number of reaches, ND (m) &
                = Time step, DELT (days) &
                = '/
      DATA TITLE(21)/' Number of stations, NS &
                = Duration of simulation, TFIN (jours) &
                = '/
      DATA TITLE(22)/' '/
      DATA TITLE(23)/' PRINTING OF RESULTS: '/
      DATA TITLE(24)/' Printing of the results every NPP step &
                = '/
      DATA TITLE(25)/' Name of the output file (max. 40 char.) &
                = '/
      WRITE(NOUT,10)
10    FORMAT(//' PROGRAM FOR BED-LOAD SEDIMENT TRANSPORT CALC&
            ULATION'/&
            ' BY TAKING INTO ACCOUNT BED-LEVEL MODIFICAT&
            IONS.'//&
            ' NOTES :'/' - UNIT SYSTEM = SI'/&
            ' - NUM. METHOD FOR WATER-SURFACE PROFILE CALCULATION = 4&
                  th ORDER RUNGE-KUTTA'/&
            ' - THE FLOW IS SUBCRITICAL (Fr < 1). THE WATER SURFACE &
                  PROFILE CALCULATION'/&
            ' STARTS AT THE DOWNSTREAM END AND PROGRESSES TOWARDS T&
                  HE UPSTREAM END'/&
            ' - SEDIMENT TRANSPORT CALCULATIONS ARE CARRIED OUT FROM &
                  UPSTREAM TO DOWNSTREAM'/&
            ' - CALCULATIONS ARE MADE FOR A UNIT WIDTH'//)
      WRITE(TITLE(2)(53:61),'(F9.7)') SF
      WRITE(TITLE(3)(53:56),'(F4.2)') D50*1000.0
      WRITE(TITLE(4)(53:58),'(F6.4)') CN50
      WRITE(TITLE(5)(53:58),'(F6.4)') CN
      WRITE(TITLE(6)(53:59),'(F7.2)') ROS
      WRITE(TITLE(7)(53:59),'(F7.2)') ROE
      WRITE(TITLE(8)(53:60),'(F8.2)') QU
      WRITE(TITLE(11)(53:57),'(F5.3)') DHDYNM
      WRITE(TITLE(11)(119:122),'(F4.2)') VARZMX
      WRITE(TITLE(12)(53:53),'(I1)') NMC
      WRITE(TITLE(12)(119:124),'(F6.4)') CFOI
      WRITE(TITLE(13)(119:123),'(F5.3)') HAMHAV
      WRITE(TITLE(16)(53:61),'(F9.2)') X1
      WRITE(TITLE(17)(53:61),'(F9.2)') XF
      WRITE(TITLE(17)(119:125),'(F7.3)') H1
      WRITE(TITLE(18)(53:61),'(F9.2)') TL
      WRITE(TITLE(19)(53:61),'(F9.2)') DX
      WRITE(TITLE(20)(53:57),'(I5)') ND
      TEMP = DELT / 86400.0
```

```
      WRITE(TITLE(20)(119:128),'(F10.3)') TEMP
      WRITE(TITLE(21)(53:57),'(I5)') NS
      TEMP = TFIN / 86400.0
      WRITE(TITLE(21)(119:128),'(F10.3)') TEMP
      WRITE(TITLE(24)(53:57),'(I5)') NPP
      TITLE(25)(53:92) = FICHS
      IF(NFTS.EQ.1)THEN
      TITLE(2)(68:131) = 'Bed-load transport formula by Schoklitsch (1950) is
         used. '
      ELSE IF(NFTS.EQ.2)THEN
      TITLE(2)(68:131) = 'Bed-load transport formula by Meyer-Peter et all.
         (1948) is used.'
      TITLE(3)(68:118) = 'Roughness parameter, FCOR = (CN50 / CN)^(3/2) = '
      WRITE(TITLE(3)(119:125),'(F7.3)') FCOR
      ELSE IF(NFTS.EQ.3)THEN
      TITLE(2)(68:131) = 'Bed-load transport formula by Einstein (1942) is
         used. '
      ELSE IF(NFTS.EQ.4)THEN
      ENDIF
      DO 110 I = 1, 25
      WRITE(NOUT,100)TITLE(I)
100   FORMAT(A)
110   CONTINUE
      RETURN
      END


      SUBROUTINE DWRITE(NOUT, NS, T, X, ZF, ZFI, H, QU, G,&
                QSU, DELQS, DQSDET, DQSERT, CPTT,&
                DELZ, DELZMX, IMAXZ, DELZRM, IMAXR)

! HEADER (20) = LINES OF THE TITLE
! FRN        = FROUDE NUMBER AT A STATION
! TH         = TIME IN HOURS
! U          = AVERAGE VELOCITY AT A STATION
      CHARACTER*131 HEADER(9)
      DOUBLE PRECISION T, ZF (NS), ZFI (NS)
      DOUBLE PRECISION QSU(NS), DELQS(NS), DQSDET, DQSERT
      DOUBLE PRECISION DELZ(NS), DELZMX, DELZRM
      DIMENSION X(NS), H(NS)
      HEADER(1) =' Time step = Total vol&
      ume of deposited sediments, DQSDET (m3/m): '
      HEADER(2) =' Time (s) = Total vol&
      ume of eroded sediments, DQSERT (m3/m): '
      HEADER(3) =' Time (hours) = '
      HEADER(4) =' Time (days) = Max. absolute be&
      d-level variation, DELZMX (m) = at the station, IMA&
      XZ = '
      HEADER(5) =' Time (years) = Max. relative be&
      d-level variation, DELZRM (-) = at the station, IMA&
      XR = '
      HEADER(6) =' '
      HEADER(7) =' '
      HEADER(8) =' STATION X ZF ZF - ZFI H &
         ZF + H U Fr QSU DELQS &
      DELZ'
      HEADER(9) =' NO (m) (m) (m) (m)&
```

```
         (m)  (m/s)  (-)  (m3/s/m)  (m3/s/m)  &
   (m)'

   WRITE(HEADER(1)(19:28),'(F10.0)')CPTT
   WRITE(HEADER(2)(19:33),'(D15.7)')T
   TH = T / 3600.0
   WRITE(HEADER(3)(19:28),'(F10.3)')TH
   TJ = T / 86400.0
   WRITE(HEADER(4)(19:28),'(F10.3)')TJ
   TA = T / (86400.0 * 365)
   WRITE(HEADER(5)(19:28),'(F10.3)')TA
   WRITE(HEADER(1)(101:112),'(D12.6)')DQSDET
   WRITE(HEADER(2)(101:112),'(D12.6)')DQSERT
   WRITE(HEADER(4)(89:100),'(D12.6)')DELZMX
   WRITE(HEADER(4)(124:127),'(I4)')IMAXZ
   WRITE(HEADER(5)(89:100),'(D12.6)')DELZRM
   WRITE(HEADER(5)(124:127),'(I4)')IMAXR
! WRITE THE RESULTS FOR THE RUNNING TIME
   WRITE(NOUT,10)
10    FORMAT(//1X,' ================================================&
   ==========<0>=======================================================&
   ============'/)
   DO 30 I = 1, 9
   WRITE(NOUT,20)HEADER(I)
20    FORMAT(A)
30    CONTINUE
   DO 50 I = 1, NS
   U = QU / H(I)
   IF(I.EQ.1) U = 0.0
   FRN = U / SQRT(G * H(I))
   WRITE(NOUT,40) I, X(I), ZF(I), ZF(I)-ZFI(I), H(I),&
   ZF(I)+H(I), U, FRN, QSU(I), DELQS(I), DELZ(I)
40    FORMAT(3X, I4, 3X, F9.2, 2X, D12.6, 2X, D12.6, 2X, F7.3,&
         2X, D12.6, 2X, F6.3, 2X, F6.3, 2x, D12.6, 2X,&
         D12.6, 2X, D12.6)
50    CONTINUE
   WRITE(8,100)TA
100   FORMAT('TIME (years) = ,',F10.3)
   DO 120 I = 1, NS
   WRITE(8,110) I, X(I), ZF(I), ZF(I)+H(I)
110   FORMAT(I4 , ',' , F9.2 , ',' D12.6 , ',' , D12.6)
120   CONTINUE
   WRITE(8,130)
130   FORMAT(/)
   RETURN
   END

!1) COMPUTER PROGRAM CODE OF SUBROUTINE SCHOKL
SUBROUTINE SCHOKL(QU, D50, SS, SEFF, QCRIT, QSU)

! QCRIT= CRITICAL DISCHARGE
   DOUBLE PRECISION QSU
   QCRIT = 0.26 * (SS - 1)**(5./3.) * D50**1.5 / SEFF**(7./6.)
   IF(QU-QCRIT.LE.0)THEN
     QSU = 0.0
   ELSE
```

```
      QSU = 2.5 * SEFF**1.5 * (QU - QCRIT) / SS
   ENDIF
   RETURN
   END
```

!2) COMPUTER PROGRAM CODE OF SUBROUTINE MEYPET

```
SUBROUTINE MEYPET(G, SEFF, ROE, ROS, D50, FCOR, RH, QSU)

! RH = HYDRAULIC RADIUS (= H)
! TERM1 = NOMINATOR IN MEYER-PETER'S FORMULA
! TERM2 = DENOMINATOR IN MEYER-PETER'S FORMULA
   DOUBLE PRECISION TERM1, TERM2, QSU
   TERM1 = (G * ROE * RH * FCOR * SEFF) / D50
   TERM1 = TERM1 - (0.047 * G * (ROS - ROE))
   IF(TERM1.LE.0.0)THEN
    QSU = 0.0
    RETURN
   ENDIF
   TERM2 = D50 / (0.25 * ROE**(1./3.))
   QSU = (TERM2 * TERM1)**(3./2.) / (G * (ROS - ROE))
   RETURN
   END
```

!3) COMPUTER PROGRAM CODE OF SUBROUTINE EINS42

```
SUBROUTINE EINS42(SS, G, D50, RH, SEFF, QSU)
! RH= HYDRAULIC RADIUS (= H)
! TERM1= TERM MULTIPLYING THE EXPONENTIAL TERM
! TERM2= EXPONENTIAL TERM
   DOUBLE PRECISION TERM1, TERM2, QSU
! COMPUTE THE BED-LOAD DISCHARGE
   TERM1 = SQRT((SS - 1.0) * G * D50**3) / 0.465
   TERM2 = 0.391 * (SS - 1.0) * D50 / (RH * SEFF)
   QSU = TERM1 * EXP(-TERM2)
   RETURN
   END
```

!4) COMPUTER PROGRAM CODE OF SUBROUTINE INGLAC

```
SUBROUTINE INGLAC(G, D50, SS, QU, H, QSU)
! W : FALL VELOCITY OF SEDIMENT PARTICLE
! KVIS;KINEMATIC VISCOSITY OF WATER
! V :AVERAGE FLOW VELOCITY
! A :FIRST TERM OF THE RUBEY (1933) FALL VELOCITY FORMULA
! B :SECOND TERM OF THE RUBEY (1933) FALL VELOCITY FORMULA
! F :DIFFERENCE OF TERM1 AND TERM2 OF RUBEY'S FORMULA

   REAL W, V
   DOUBLE PRECISION KVIS, A, B, F, QSU
! CALCULATION OF FALL VELOCITY
   KVIS=0.000001
   A=(0.667+36*KVIS**2/(G*D50**3*(SS-1)))**0.5
   B=(36*KVIS**2/(G*D50**3*(SS-1)))**0.5
   IF (D50.EQ.0.001.OR.D50.GT.0.001) F=0.79
   IF (D50.LT.0.001) F=A-B
```

```
      W=F*(D50*G*(SS-1))**0.5
! CALCULATION OF SEDIMENT DISCHARGE
      V=QU/H
      QSU=0.562*(KVIS*G)**0.3333*V**5/(W*G*H*G*SS)
      PRINT*,"SEDIMENT DISCHARGE VOLUME",QSU,"m3/s/m"
      RETURN
      END
```

!5) COMPUTER PROGRAM CODE OF SUBROUTINE ENGEHAN

```
SUBROUTINE ENGEHAN(ROE, G, H, SEFF, QU, SF, SS, D50, QSU)

! FP = DIMENSIONLESS PARAMETER IN THE PROCEDURE
! THETA =DIMENSIONLESS SHEAR STRESS PARAMTER OF SHIELDS
! V = AVERAGE FLOW VELOCITY
! PHI = PARAMETER USED IN THE PROCEDURE
! SHE = BED SHEAR STRESS

      REAL V,FP,THETA,PHI
      DOUBLE PRECISION SHE,QSU
      SHE=ROE*G*H*SEFF
      V=QU/H
      FP=2*G*SF*H/V**2
      THETA=SHE/((ROE*SS*G-ROE*G)*D50)
      PHI=0.1*THETA**2.5/FP
      QSU=PHI*((SS-1)*G*D50**3)**0.5
      PRINT*,"SEDIMENT DISCHARGE=",QSU,"m3/s/m"
      RETURN
      END
```

!6) COMPUTER PROGRAM CODE OF SUBROUTINE DUBOY
```
SUBROUTINE DUBOY(ROE, G, H, SEFF, D50, SS, QSU)

! KVIS=KINEMATIC VISCOSITY OF WATER
! SHE= BED SHEAR STRESS
! CSHE= CRITICAL BED SHEAR STRESS
! F= VANONI'S (1977) PARAMETER TO DETERMINE CRITICAL SHEAR
!    STRESS
! A=DIMENSIONLESS SHIELDS PARAETER OF CORRESPONDING F
!    VALUE

!
      REAL KVIS,A,F
      DOUBLE PRECISION QSU,SHE,CSHE
      KVIS=0.000001

      SHE=ROE*G*H*SEFF
      SHE=SHE*0.2248/3.281**2

! CALCULATION OF CRITICAL SHEAR STRESS (CSHE)
      F=(D50/(KVIS))*(0.1*(SS-1)*G*D50)**0.5

      IF(F.EQ.500.OR.F.GT.500) A=0.06
        IF(F.LT.500.AND.F.GT.400.OR.F.EQ.400) A=0.057
        IF(F.LT.400.AND.F.GT.300.OR.F.EQ.300) A=0.055
      IF(F.LT.300.AND.F.GT.200.OR.F.EQ.200) A=0.050
```

```
        IF(F.LT.200.AND.F.GT.150.OR.F.EQ.150) A=0.045
        IF(F.LT.150.AND.F.GT.100.OR.F.EQ.100) A=0.040
        IF(F.LT.100.AND.F.GT.90.OR.F.EQ.90) A=0.039
        IF(F.LT.90.AND.F.GT.80.OR.F.EQ.80) A=0.038
        IF(F.LT.80.AND.F.GT.70.OR.F.EQ.70) A=0.037
        IF(F.LT.70.AND.F.GT.60.OR.F.EQ.60) A=0.035
        IF(F.LT.60.AND.F.GT.50.OR.F.EQ.50) A=0.035
        IF(F.LT.50.AND.F.GT.40.OR.F.EQ.40) A=0.033
        IF(F.LT.40.AND.F.GT.30.OR.F.EQ.30) A=0.032
        IF(F.LT.30.AND.F.GT.20.OR.F.EQ.20) A=0.031
        IF(F.LT.20.AND.F.GT.15.OR.F.EQ.15) A=0.031
        IF(F.LT.15.AND.F.GT.10.OR.F.EQ.10) A=0.035
        IF(F.LT.10.AND.F.GT.9.OR.F.EQ.9) A=0.035
        IF(F.LT.9.AND.F.GT.8.OR.F.EQ.8) A=0.036
        IF(F.LT.8.AND.F.GT.7.OR.F.EQ.7) A=0.038
        IF(F.LT.7.AND.F.GT.6.OR.F.EQ.6) A=0.036
     IF(F.LT.6.AND.F.GT.5.OR.F.EQ.5) A=0.045
        IF(F.LT.5.AND.F.GT.4.OR.F.EQ.4) A=0.050
        IF(F.LT.4.AND.F.GT.3.OR.F.EQ.3) A=0.058
        IF(F.LT.3.AND.F.GT.2.OR.F.EQ.2) A=0.070
        IF(F.LT.2.AND.F.GT.1.OR.F.EQ.1) A=0.082


     CSHE=(SS*ROE*G-ROE*G)*D50*A
     CSHE=CSHE*4.448/3.281**2


     IF (SHE.LT.CSHE) THEN
       PRINT*,"NO SEDIMENT TRANSPORT SINCE BED-SHEAR IS LESS THEN CRITICAL
              SHEAR STRESS"
     GO TO 1
     END IF
!    QSU=(0.173/(D50*1000)**0.75)*SHE*(SHE-CSHE)
     QSU=QSU*0.3048**2

1    CONTINUE

     RETURN
     END


!7) COMPUTER PROGRAM CODE OF SUBROUTINE KALIN

SUBROUTINE KALIN(ROE, G, H, SEFF, D50, SS, QSU)

! KVIS= KINEMATIC VISCOSITY OF WATER
! SHE=BED SHEAR STRESS
! CSHE= CRITICAL BED SHEAR STRESS
! F=VANONI'S (1977) PARAMETER TO DETERMINE CRITICAL SHEAR STRESS
! A= DIMENSIONLESS SHIELDS PARAETER OF CORRESPONDING F
!    VALUE
! RATIO=RATIO OF SHE TO CSHE
! USTAR= SHEAR VELOCITY
! F2= FIRST TERM USED IN THE PROCEDURE
! F3= SECOND TERM USED IN THE PROCEDURE
! F4= SUM OF F2 AND F3
! F5= THIRD TERM USED IN THE PROCEDURE
```

```
        REAL SHE,F,KVIS,A,CSHE,RATIO
        REAL USTAR
        DOUBLE PRECISION QSU,F3,F4,F2,F5
        KVIS=0.000001
        SHE=ROE*G*H*SEFF

        F=(D50/(KVIS))*(0.1*(SS-1)*G*D50)**0.5
                IF(F.EQ.500.OR.F.GT.500) A=0.06
           IF(F.LT.500.AND.F.GT.400.OR.F.EQ.400) A=0.057
           IF(F.LT.400.AND.F.GT.300.OR.F.EQ.300) A=0.055
        IF(F.LT.300.AND.F.GT.200.OR.F.EQ.200) A=0.050
        IF(F.LT.200.AND.F.GT.150.OR.F.EQ.150) A=0.045
        IF(F.LT.150.AND.F.GT.100.OR.F.EQ.100) A=0.040
        IF(F.LT.100.AND.F.GT.90.OR.F.EQ.90) A=0.039
        IF(F.LT.90.AND.F.GT.80.OR.F.EQ.80) A=0.038
        IF(F.LT.80.AND.F.GT.70.OR.F.EQ.70) A=0.037
        IF(F.LT.70.AND.F.GT.60.OR.F.EQ.60) A=0.035
        IF(F.LT.60.AND.F.GT.50.OR.F.EQ.50) A=0.035
        IF(F.LT.50.AND.F.GT.40.OR.F.EQ.40) A=0.033
        IF(F.LT.40.AND.F.GT.30.OR.F.EQ.30) A=0.032
        IF(F.LT.30.AND.F.GT.20.OR.F.EQ.20) A=0.031
        IF(F.LT.20.AND.F.GT.15.OR.F.EQ.15) A=0.031
        IF(F.LT.15.AND.F.GT.10.OR.F.EQ.10) A=0.035
        IF(F.LT.10.AND.F.GT.9.OR.F.EQ.9) A=0.035
        IF(F.LT.9.AND.F.GT.8.OR.F.EQ.8) A=0.036
        IF(F.LT.8.AND.F.GT.7.OR.F.EQ.7) A=0.038
        IF(F.LT.7.AND.F.GT.6.OR.F.EQ.6) A=0.036
    IF(F.LT.6.AND.F.GT.5.OR.F.EQ.5) A=0.045
        IF(F.LT.5.AND.F.GT.4.OR.F.EQ.4) A=0.050
        IF(F.LT.4.AND.F.GT.3.OR.F.EQ.3) A=0.058
        IF(F.LT.3.AND.F.GT.2.OR.F.EQ.2) A=0.070
        IF(F.LT.2.AND.F.GT.1.OR.F.EQ.1) A=0.082

        CSHE=(SS*ROE*G-ROE*G)*D50*A
        RATIO=CSHE/SHE

        F2=-0.068-1.1328*RATIO+0.94*RATIO**2.0-1.206*RATIO**3.0
        F3=0.567*RATIO**4.0-0.0975*RATIO**5.0
        F4=F2+F3
        F5=10.0**F4

        USTAR=(SHE/ROE)**0.5
        QSU=USTAR*D50*F5
        PRINT*,"SEDIMENT DISCHARGE (m3/s/m)",QSU

        RETURN
        END

!8) COMPUTER PROGRAM CODE OF SUBROUTINE SHIELD

SUBROUTINE SHIELD (G,ROE,H,SEFF,D50,SS,QU,SF,QSU)
!KVIS = KINEMATIC VISCOSITY OF WATER
!SHE = BED SHEAR STRESS
!CSHE = CRITICAL BED SHEAR STRESS
!F = VANONI'S (1977) PARAMETER TO DETERMINE CRITICAL SHEAR
!     STRESS
```

```
!A= DIMENSIONLESS SHIELDS PARAMETER OF CORRESPONDING F
!    VALUE

      REAL SHE,CSHE,KVIS,F,A
       DOUBLE PRECISION QSU
       KVIS=0.000001
          SHE=G*ROE*H*SEFF
       F=(D50/(KVIS))*(0.1*(SS-1)*G*D50)**0.5
       PRINT*,"VALUE OF F IS",F
       IF(F.EQ.500.OR.F.GT.500) A=0.06
          IF(F.LT.500.AND.F.GT.400.OR.F.EQ.400) A=0.057
          IF(F.LT.400.AND.F.GT.300.OR.F.EQ.300) A=0.055
       IF(F.LT.300.AND.F.GT.200.OR.F.EQ.200) A=0.050
       IF(F.LT.200.AND.F.GT.150.OR.F.EQ.150) A=0.045
       IF(F.LT.150.AND.F.GT.100.OR.F.EQ.100) A=0.040
       IF(F.LT.100.AND.F.GT.90.OR.F.EQ.90) A=0.039
       IF(F.LT.90.AND.F.GT.80.OR.F.EQ.80) A=0.038
       IF(F.LT.80.AND.F.GT.70.OR.F.EQ.70) A=0.037
       IF(F.LT.70.AND.F.GT.60.OR.F.EQ.60) A=0.035
       IF(F.LT.60.AND.F.GT.50.OR.F.EQ.50) A=0.035
       IF(F.LT.50.AND.F.GT.40.OR.F.EQ.40) A=0.033
       IF(F.LT.40.AND.F.GT.30.OR.F.EQ.30) A=0.032
       IF(F.LT.30.AND.F.GT.20.OR.F.EQ.20) A=0.031
       IF(F.LT.20.AND.F.GT.15.OR.F.EQ.15) A=0.031
       IF(F.LT.15.AND.F.GT.10.OR.F.EQ.10) A=0.035
       IF(F.LT.10.AND.F.GT.9.OR.F.EQ.9) A=0.035
       IF(F.LT.9.AND.F.GT.8.OR.F.EQ.8) A=0.036
       IF(F.LT.8.AND.F.GT.7.OR.F.EQ.7) A=0.038
       IF(F.LT.7.AND.F.GT.6.OR.F.EQ.6) A=0.036
        IF(F.LT.6.AND.F.GT.5.OR.F.EQ.5) A=0.045
       IF(F.LT.5.AND.F.GT.4.OR.F.EQ.4) A=0.050
       IF(F.LT.4.AND.F.GT.3.OR.F.EQ.3) A=0.058
       IF(F.LT.3.AND.F.GT.2.OR.F.EQ.2) A=0.070
       IF(F.LT.2.AND.F.GT.1.OR.F.EQ.1) A=0.082

      CSHE=(SS*ROE*G-ROE*G)*D50*A

      IF (SHE.LT.CSHE) THEN
      GO TO 1
      END IF

QSU=QU*(SF/SS)*10*(SHE-CSHE)/((H*(SS*ROE*G-ROE*G)))

1     CONTINUE

      RETURN
      END

!9) COMPUTER PROGRAM CODE OF SUBROUTINE LAURSEN

SUBROUTINE LAURSEN(G,D50,SS,QU,H,ROE,QSU)

! KVIS=KINEMATIC VISCOSITY OF WATER
! GRSHE= SHEAR STRESS DUE TO GRAIN ROUGHNESS
! CSHE= CRITICAL BED SHEAR STRESS
! F= VANONI'S (1977) PARAMETER TO DETERMINE CRITICAL SHEAR
```

```
!      STRESS
! A =DIMENSIONLESS SHIELDS PARAETER OF CORRESPONDING F
!      VALUE
! W = FALL VELOCITY
! USTAR= FRICTION VELOCITY
! V = AVERAGE FLOW VELOCITY
! CT= TOTAL AVERAGE SEDIMENT CONCENTRATION BY WEIGHT
! FSED = FACTOR FOR THE EFFECT OF SUSPENDED SEDIMENT
! RATIO2=RATIO OF RGSHE/CSHE
! K =FIRST TERM OF RUBEYS'S FALL VELOCITY FORMULA
! U =CONSTANT (=079) IN RUBEY'S FALL VELOCITY FORMULA
! B =SECOND TERM OF RUBEYS'S FALL VELOCITY FORMULA

       REAL W,V,FSED,RATIO2
        DOUBLE PRECISION QSU,KVIS,A,B,F,U,K
        DOUBLE PRECISION GRSHE,CSHE
! CALCULATION OF FALL VELOCITY
   KVIS=0.000001

    K=(0.667+36*KVIS**2/(G*D50**3*(SS-1)))**0.5
    B=(36*KVIS**2/(G*D50**3*(SS-1)))**0.5

    IF (D50.EQ.0.001.OR.D50.GT.0.001) U=0.79
    IF (D50.LT.0.001) U=K-B

    W=U*(D50*G*(SS-1))**0.5

! CALCULATION OF SHEAR STRESS DUE TO GRAIN ROUGHNESS
    V=QU/H
    GRSHE=(ROE*V**2/58)*(D50/H)**0.3333

! CALCULATION OF CRITICAL SHEAR STRESS (CSHE)

    F=(D50/(KVIS))*(0.1*(SS-1)*G*D50)**0.5

    IF(F.EQ.500.OR.F.GT.500) A=0.06
       IF(F.LT.500.AND.F.GT.400.OR.F.EQ.400) A=0.057
       IF(F.LT.400.AND.F.GT.300.OR.F.EQ.300) A=0.055
    IF(F.LT.300.AND.F.GT.200.OR.F.EQ.200) A=0.050
    IF(F.LT.200.AND.F.GT.150.OR.F.EQ.150) A=0.045
    IF(F.LT.150.AND.F.GT.100.OR.F.EQ.100) A=0.040
    IF(F.LT.100.AND.F.GT.90.OR.F.EQ.90) A=0.039
    IF(F.LT.90.AND.F.GT.80.OR.F.EQ.80) A=0.038
    IF(F.LT.80.AND.F.GT.70.OR.F.EQ.70) A=0.037
    IF(F.LT.70.AND.F.GT.60.OR.F.EQ.60) A=0.035
    IF(F.LT.60.AND.F.GT.50.OR.F.EQ.50) A=0.035
    IF(F.LT.50.AND.F.GT.40.OR.F.EQ.40) A=0.033
    IF(F.LT.40.AND.F.GT.30.OR.F.EQ.30) A=0.032
    IF(F.LT.30.AND.F.GT.20.OR.F.EQ.20) A=0.031
    IF(F.LT.20.AND.F.GT.15.OR.F.EQ.15) A=0.031
    IF(F.LT.15.AND.F.GT.10.OR.F.EQ.10) A=0.035
    IF(F.LT.10.AND.F.GT.9.OR.F.EQ.9) A=0.035
    IF(F.LT.9.AND.F.GT.8.OR.F.EQ.8) A=0.036
    IF(F.LT.8.AND.F.GT.7.OR.F.EQ.7) A=0.038
    IF(F.LT.7.AND.F.GT.6.OR.F.EQ.6) A=0.036
       IF(F.LT.6.AND.F.GT.5.OR.F.EQ.5) A=0.045
```

```
      IF(F.LT.5.AND.F.GT.4.OR.F.EQ.4) A=0.050
      IF(F.LT.4.AND.F.GT.3.OR.F.EQ.3) A=0.058
      IF(F.LT.3.AND.F.GT.2.OR.F.EQ.2) A=0.070
      IF(F.LT.2.AND.F.GT.1.OR.F.EQ.1) A=0.082

      CSHE=(SS*ROE*G-ROE*G)*D50*A
      RATIO2=GRSHE/CSHE

      IF (RATIO2.LT.1.OR.RATIO2.EQ.1) THEN

      GOTO 1
      ENDIF

!C CALCULATION OF SEDIMENT DISCHARGE
!C NO EEFFECT OF SUSPENDED SEDIMENT FSED=1

      FSED=1
      QSU=QU*(D50/H)**1.66667*(GRSHE/CSHE-1)*FSED

1     CONTINUE
      RETURN
      END

!10) COMPUTER PROGRAM CODE OF SUBROUTINE ACKWHI

SUBROUTINE ACKWHI (QU,H,G,D50,SS,ROE,SEFF,QSU)

! SUBROUTINE SUBPROGRAM FOR CALCULATING THE TOTAL-LOAD TRANSPORT AT
! KVIS=KINEMATIC VISCOSITY OF WATER
! SHE= BED SHEAR STRESS
! CSHE=CRITICAL BED SHEAR STRESS
! N= TRANSITION EXPONENT, DEPENDING ON SEDIMENT SIZE
! A= PARAMETER USED IN THE CALCULATION OF GENERAL
!   DIMENSIONLESS SEDIMENT TRANSPORT NUMBER
! M= PARAMETER USED IN THE CALCULATION OF GENERAL
!   DIMENSIONLESS SEDIMENT TRANSPORT NUMBER
! C= PARAMETER USED IN THE CALCULATION OF GENERAL
!   DIMENSIONLESS
!             SEDIMENT TRANSPORT NUMBER
! LC= EXPONENT VALUE IN ORDER TO CALCULATE C PARAMETER
! USTAR= FRICTION VELOCITY
! V=AVERAGE FLOW VELOCITY
! CT= TOTAL AVERAGE SEDIMENT CONCENTRATION BY WEIGHT
! ALPHA= COEFFICIENT IN ROUGH TURBULENT FLOW (=10)
! DGR = DIMENSIONLESS GRAIN DIAMETER
! FGR = DIMENSIONLESS MOBILITY NUMBER
! GGR = GENERAL DIMENSIONLESS SEDIMENT TRANSPORT FUNCTION
! RATE = RATE OF SEDIMENT TRANSPORT IN TERMS OF MASS FLOW PER
!    UNIT MASS FLOW RATE
!

      REAL FR,KVIS,USTAR,SHE,A,N,M,C,ALPHA,LC
      DOUBLE PRECISION DGR,FGR,GGR,RATIO,RATE,QSU

      KVIS=0.000001
! CALCULATION OF FROUDE NUMBER
```

```
      V=QU/H
      FR=V**2/(G*H)

      IF (FR.GT.0.8) THEN
      PRINT*,"FROUDE NUMBER IS NOT PROPER FOR CALCULATION OF PARAMETERS"
      GOTO 2
      END IF

! CALCULATION OF DIMENSIONLESS GRAIN DIAMETER
      DGR=D50*(G*(SS-1)/KVIS**2)**0.3333
! CALCULATION OF BED-SHEAR SRESS (N/m2)
      SHE=ROE*G*H*SEFF

! CALCULATION OF SHEAR VELOCITY, USTAR
      USTAR=(SHE/ROE)**0.5
! CALCULATION OF A,n,m,C PARAMETERS
      IF (DGR.GT.1.AND.DGR.LT.60.OR.DGR.EQ.60) THEN
      N=1-0.56*LOG10(DGR)
      A=0.23/DGR**0.5+0.14
      M=9.66/DGR+1.34
      LC=2.86*LOG10(DGR)-(LOG10(DGR))**2-3.53
      IF (LC.LT.0) THEN
      C=0.1**(-LC)
      GO TO 1
      END IF
      C=10**LC
1     END IF
      IF (DGR.GT.60) THEN
      N=0
      A=0.17
      M=1.5
      C=0.025
      END IF

      ALPHA=10
      FGR=(USTAR**N/(G*D50*(SS-1))**0.5)*(V/(32**0.5*LOG10(ALPHA*H/D50)))&
      )**(1-N)

      RATIO=FGR/A
      IF (RATIO.LT.1) THEN

      GO TO 2
      END IF

      GGR=C*(FGR/A-1)**M
      RATE=GGR*D50*SS/H*(V/USTAR)**N

      QSU=QU*RATE/SS
2     CONTINUE

      RETURN
      END
!11) COMPUTER PROGRAM CODE OF SUBROUTINE SHUNG

SUBROUTINE SHUNG(G,D50,SS,QU,H,SF,QSU)
```

```
!
! KVIS=KINEMATIC VISCOSITY OF WATER
! V= AVERAGE FLOW VELOCITY
! W= FALL VELOCITY OF SEDIMENT PARTICLES
! K= FIRST TERM OF RUBEYS'S FALL VELOCITY FORMULA
! U= CONSTANT (=079) IN RUBEY'S FALL VELOCITY FORMULA
! B= SECOND TERM OF RUBEYS'S FALL VELOCITY FORMULA
! Y= PARAMETER WH?CH DEPENDS ON V, W AND RIVER SLOPE (SF)
! CT= SEDIMENT CONCENTRATION IN ppm BY WEIGHT

      REAL W,V
       DOUBLE PRECISION KVIS,B,U,K,Y,CT,QSU!
! CALCULATION OF FALL VELOCITY
      KVIS=0.000001

      K=(0.667+36*KVIS**2/(G*D50**3*(SS-1)))**0.5
      B=(36*KVIS**2/(G*D50**3*(SS-1)))**0.5
      IF (D50.EQ.0.001.OR.D50.GT.0.001) U=0.79
      IF (D50.LT.0.001) U=K-B
      W=U*(D50*G*(SS-1))**0.5
! UNIT CONVERSION
      W=3.281*W
! CALCULATION OF AVERAGE FLOW VELOCITY
      V=QU/H
! UNIT CONVERSION
      V=3.281*V
! CALCULATION O PARAMETER Y
      Y=(V*SF**0.57/W**0.32)**0.00750189
! CALCULATION OF CT
      CT=-107404.45938164+324214.74734085*Y-326309.58908739*Y**2+109503 &
             *.87232539*Y**3
             CT=10**CT
! CALCULATION OF QSU QSU=q*CT/10**6
             QSU=QU*CT/10**6

      RETURN
      END


!12) COMPUTER PROGRAM CODE OF SUBROUTINE YANG

SUBROUTINE YANG (G,D50,SS,QU,H,ROE,SEFF,QSU)
!
!KVIS= KINEMATIC VISCOSITY OF WATER
!W= FALL VELOCITY OF SEDIMENT PARTICLES
!K= FIRST TERM OF RUBEYS'S FALL VELOCITY FORMULA
!U= CONSTANT (=079) IN RUBEY'S FALL VELOCITY FORMULA
!B= SECOND TERM OF RUBEYS'S FALL VELOCITY FORMULA
!VCR=CRITICAL VELOCITY
!SHE=BED SHEAR STRESS
!USTAR= SHEAR VELOCITY
!RATIO1= VCR/W
!SRE= SHEAR REYNOLDS NUMBER DEFINED AS USTAR*D50/KVIS
!TERM1= FIRST TERM USED IN THE CALCULATION OF SEDIMENT DISCAHRGE
!TERM2= SECOND TERM USED IN THE CALCULATION OF SEDIMENT ISCAHRGE
!AX = CHECK PARAMETER FOR SEDIMENT DISCHARGE CALCULATION
```

```
!MFW=     MASS FLOW RATE
!CTS=     TOTAL SAND CONCENTRATION IN ppm by Weight
!

    REAL SHE,USTAR,RATIO1,V,MFW
    DOUBLE PRECISION KVIS,B,U,K,AX
    DOUBLE PRECISION QSU,SRE,TERM1,TERM2,CTS
!C CALCULATION OF FALL VELOCITY

    KVIS=0.000001

    K=(0.667+36*KVIS**2/(G*D50**3*(SS-1)))**0.5
    B=(36*KVIS**2/(G*D50**3*(SS-1)))**0.5

    IF (D50.EQ.0.001.OR.D50.GT.0.001) U=0.79
    IF (D50.LT.0.001) U=K-B

    W=U*(D50*G*(SS-1))**0.5
!C CALCULATION OV AVERAGE FLOW VELOCITY
    V=QU/H
!C CALCULATION OF BED-SHEAR SRESS (N/m2)
    SHE=ROE*G*H*SEFF
!C CALCULATION OF SHEAR VELOCITY, USTAR
    USTAR=(SHE/ROE)**0.5
!C CALCULATION OF SEDIMENT DISCHARGE
    SRE=USTAR*D50/KVIS
            IF (SRE.GT.1.2.AND.SRE.LT.70.OR.SRE.EQ.1.2) THEN
    RATIO1=2.5/(LOG10(SRE)-0.06)+0.66
    END IF

    IF (SRE.GT.70.OR.SRE.EQ.70) THEN
    RATIO1=2.05
    END IF

    IF (SRE.LT.1.2) THEN
    GO TO 1
    END IF

    TERM1=5.435-0.286*LOG10(W*D50/KVIS)-0.457*LOG10(USTAR/W)

    AX=V*SEFF/W-RATIO1*SEFF
    IF (AX.LT.0.OR.AX.EQ.0) THEN
    GOTO 1
    END IF

    TERM2=(1.799-0.409*LOG10(W*D50/KVIS)-0.314*LOG10(USTAR/W))
            *LOG10(V**SEFF/W-RATIO1*SEFF)

    CTS=TERM1+TERM2
    CTS=10**CTS

    MFW=QU*ROE
    QSU=MFW*CTS/(1000000*SS*ROE)

1   CONTINUE
```

```
      RETURN
      END
```

!13) COMPUTER PROGRAM CODE OF SUBROUTINE ROTTNER

```
SUBROUTINE ROTTNER (QU,H,D50,SS,G,QSU)
!
! V= AVERAGE FLOW VELOCITY
! TERM1=FIRST TERM USED IN THE CALCULATION PROCEDURE
! TERM2=SECOND TERM USED IN THE CALCULATION PROCEDURE
! TERM3=THIRD TERM USED IN THE CALCULATION PROCEDURE
! DECLERATION OF VARIABLES
      DOUBLE PRECISION QSU,TERM1,TERM2,TERM3
      REAL V

! CALCULATION OF AVERAGE VELOCITY
      V=QU/H
! CALCULATION OF SEIDMENT DISCHARGE
      TERM1=((SS-1)*G*H**3)**0.5
      TERM2=V*(0.667*(D50/H)**0.6667+0.14)/((SS-1)*G*H)**0.5
      TERM3=0.778*(D50/H)**0.667
      IF (TERM2.LT.TERM3) THEN
      GOTO 1
      END IF
      QSU=TERM1*(TERM2-TERM3)**3

1     CONTINUE

      RETURN
      END
```

!14) COMPUTER PROGRAM CODE OF SUBROUTINE TOFFALETI

```
SUBROUTINE TOFFALETI (G,D50,SS,QU,H,SF,QSU)

! V= AVERAGE FLOW VELOCITY
! W= FALL VELOCITY
! KVIS= KINEMATIC VISCOSITY OF WATER
! CZ=A PARAMETER THAT DEPENDS ON THE TEMPERATURE
! Z=A PARAMETER THAT ID USED IN THE EXPONENTS
! NV=A PARAMETER THAT ID USED IN THE EXPONENTS
! N3=THIRD EXPONENT
! N2=SECOND EXPONENT
! N1=FIRST EXPONENT
! TT= TEMPERATURE PARAMETER
! GRUSTAR=SHEAR VELOCITY DEPENDING ON GRAIN ROUGHNESS
! AC= COEFFICIENT DETERMINED FROM THE GRAPH
! K = COEFFICIENT DETERMINED FROM THE GRAPH
! ACK=COEFFICIENT CALCULATED AS AC*K
! CL= SEDIMENT CONCENTRATION IN THE LOWER ZONE
! C2D50=SEDIMENT CONCENTRATION AT THE Y=2*D50
! TERM1=TERM THAT IS USED TO DETERMINE THE AC VALUE
! TERM2=TERM THAT IS USED TO DETERMINE THE K VALUE
! QSL=SEDIMENT DISCHARGE IN THE LOWER ZONE
! QSUP=SEDIMENT DISCHARGE IN THE UPPER ZONE
! QSM =SEDIMENT DISCHARGE IN THE MIDDLE ZONE
```

```
! M= PARAMETER USED IN THE CALCULATION OF SEDIMENT
!    DISCHARGES OF DIFFERENT VERTICAL ZONES
! A=FIRST TERM OF RUBEY'S FALL VELOCITY FORMULA
! B=SECOND TERM OF RUBEY'S FALL VELOCITY FORMULA
! F=CONSTANT (=079) IN RUBEY'S FALL VELOCITY FORMULA
!
! DECLERATION OF VARIABLES

      REAL W,V,CZ,Z,NV,N3,N2,N1,TT,GRUSTAR,AC,K,ACK,CL,C2D50
      DOUBLE PRECISION KVIS,A,B,F,TERM1,TERM2
      DOUBLE PRECISION QSU,QSL,QSUP,QSM,M

! CALCULATION OF FALL VELOCITY (RUBEY'S FORMULA)

      KVIS=0.000001

      A=(0.667+36*KVIS**2/(G*D50**3*(SS-1)))**0.5
      B=(36*KVIS**2/(G*D50**3*(SS-1)))**0.5

      IF (D50.EQ.0.001.OR.D50.GT.0.001) F=0.79
      IF (D50.LT.0.001) F=A-B

      W=F*(D50*G*(SS-1))**0.5

! UNIT CONVERSION TO ft/s
      W=W*3.281

! CALCULATION OF FLOW VELOCITY

      V=QU/H
! UNIT CONVERSION
      V=V*3.281

! CALCULATION OF EXPONENT Z
!UNIT CONVERSION
      H=H*3.281

! CALCULATION OF CZ
! TEMPERATURE IS ASSUMED TO BE 68 F (20 C)
      CZ=260.67-0.667*68
      Z=W*V/(CZ*H*SF)
!
! CALCULATION OF NV WATER TEMPERATURE IS ASSUMED TO BE 68 F (20 C)
      NV=0.1198+0.00048*68

      IF (Z.LT.NV) Z=1.5*NV

      N3=1+NV-0.756*Z
      N2=1+NV-Z
      N1=1+NV-1.5*Z
!
! CALCULATION OF QSL

! TEMPERATURE IS ASSUMED TO BE 68 F (20 C)
      TT=1.1*(0.051+0.00009*68)
! CALCULATION OF GRUSTAR
```

```
    GRUSTAR=V/(5.75*LOG10((H*0.3048)/D50)+6.25)
! CALCULATION OF AC AND K VALUES

    TERM1=(10**5*KVIS*3.281**2)**0.333/(10*GRUSTAR)
!
! D65 IS ASSUMED TO BE EQUAL TO D50
    TERM2=(TERM1*10**5)*D50*SF/0.3048
!
!   CALCULATION OF AC VALUE
    IF (TERM1.EQ.0.1) AC=300
     IF (TERM1.GT.0.1.AND.TERM1.LT.0.2) AC=120
    IF (TERM1.EQ.0.2) AC=100
     IF (TERM1.GT.0.2.AND.TERM1.LT.0.3) AC=65
    IF (TERM1.EQ.0.3) AC=60
    IF (TERM1.GT.0.3.AND.TERM1.LT.0.4) AC=50
    IF (TERM1.EQ.0.4) AC=40
    IF (TERM1.GT.0.4.AND.TERM1.LT.0.5) AC=35
    IF (TERM1.EQ.0.5) AC=28
    IF (TERM1.GT.0.5.AND.TERM1.LT.0.7) AC=30
    IF (TERM1.EQ.0.7) AC=32
    IF (TERM1.GT.0.7.AND.TERM1.LT.0.8) AC=40
    IF (TERM1.EQ.0.8) AC=50
    IF (TERM1.GT.0.8.AND.TERM1.LT.1.3) AC=50
    IF (TERM1.EQ.1.3) AC=48
     IF (TERM1.GT.1.3.AND.TERM1.LT.2.0) AC=100
    IF (TERM1.EQ.2.0) AC=180
    IF (TERM1.GT.2.0.AND.TERM1.LT.3.0) AC=350
     IF (TERM1.EQ.2.0) AC=500
     IF (TERM1.GT.3.0.OR.TERM1.LT.0.1) THEN
    PRINT*,"AC VALUE CANNOT BE CALCULATED"
    END IF

!   CALCULATION OF K VALUE
    IF (TERM2.GT.0.2.AND.TERM2.LT.0.3) K=1.2
     IF (TERM2.EQ.0.3) K=1.3
     IF (TERM2.GT.0.3.AND.TERM2.LT.0.35) K=1.4
     IF (TERM2.EQ.0.35) K=1.5
    IF (TERM2.GT.0.35.AND.TERM2.LT.0.4) K=1.4
    IF (TERM2.EQ.0.4) K=1.3
    IF (TERM2.GT.0.4.AND.TERM2.LT.0.5) K=1.2
    IF (TERM2.EQ.0.5) K=1.0
    IF (TERM2.GT.0.5.AND.TERM2.LT.0.6) K=0.9
    IF (TERM2.EQ.0.6) K=0.8
    IF (TERM2.GT.0.6.AND.TERM2.LT.0.7) K=0.75
    IF (TERM2.EQ.0.7) K=0.7
    IF (TERM2.GT.0.7.AND.TERM2.LT.0.8) K=0.65
    IF (TERM2.EQ.0.8) K=0.6
     IF (TERM2.GT.0.8.AND.TERM2.LT.1.0) K=0.55
    IF (TERM2.EQ.1.0) K=0.5
    IF (TERM2.GT.1.0.AND.TERM2.LT.1.7) K=0.38
    IF (TERM2.EQ.1.7) K=0.3
    IF (TERM2.GT.1.7.AND.TERM2.LT.2.0) K=0.28
    IF (TERM2.EQ.2.0) K=0.28
     IF (TERM2.LT.0.2.OR.TERM2.GT.2.0) THEN
    END IF
```

```
        ACK=AC*K

        IF (ACK.LT.16.0) THEN
        ACK=16
        END IF

        IF (D50.LT.0.088) THEN
        QSL=1.095*(V**2/(TT*ACK))**0.6
        END IF

        QSL=0.6/(TT*ACK*D50*3.281/(0.00058*V**2))**1.6667
!       CALCULATION M
        M=QSL*N3/((H/11.24)**N3-(2*D50*3.281)**N3)
!
!       CALCULATION OF QSM AND QSU

        QSM=M*(H/11.24)**(0.244*Z)*((H/2.5)**N2-(H/11.24)**N2)/N2
        QSUP=M*(H/11.24)**(0.244*Z)*(H/2.5)**(0.5*Z)*(H**N1-(H/2.5)**N1)/N1


!
!       CALCULATION OF CL
        CL=M/(43.2*(1+NV)*V*H**(0.756*Z)-NV)
!
!       CALCULATION OF C2D50
        C2D50=CL*(H/(2*D50*3.281)**(0.756*Z))

        IF (C2D50.GT.100.0) THEN
        CL=100*(2*D50*3.281/H)**(0.756*Z)
!       CALCULATION OF M IF C2D50IS GREATER THAN 100 lb/ft2
        M=43.2*CL*(1+NV)*V*H**(0.756*Z)-NV
        QSU=M*(2*D50*3.281)**N3+QSM+QSUP
!
!       UNIT CONVERSION FOR QSU
        QSU=(((QSU/(24*3600))*2000)/(62.37))*(0.03048**2)
        GOTO 1
        END IF
        QSU=M*(2*D50*3.281)**N3+QSM+QSUP
!       UNIT CONVERSION FOR QSU
        QSU=(((QSU/(24*3600))*2000)/(62.37))*(0.03048**2)
!       UNIT CONVERSION FOR H
        H=H/3.281

1       CONTINUE

        RETURN
        END

!15) COMPUTER PROGRAM CODE OF SUBROUTINE CHANG

SUBROUTINE CHANG (QU,H,ROE,G,SEFF,D50,SS,SF,QSU)

! SUBROUTINE SUBPROGRAM FOR CALCULATING THE BED-LOAD TRANSPORT AT A
! SECTION USING THE FORMULA OF CHANG, SIMONS AND RICHARDSON'S
! APPROACH C (1967)
```

```
! FOR D50=0.19,0.33,0.36,0.50,0.52,0.93mm
! LIST OF VARIABLES LOCALLY DEFINED FOR SUBROUTINE SUBPROGRAM "CHANG"
!
!! V= AVERAGE FLOW VELOCITY
! SHE= BED SHEAR STRESS
! CSHE= CRITICAL BED SHEAR STRESS
! USTAR= SHEAR VELOCITY
! KT=COEFFICIENT OF THE EQUATION

      REAL V,KVIS,F,A,USTAR
       DOUBLE PRECISION SHE,CSHE,AA,KT,QSU
! CALCULATION OF AVERAGE FLOW VELOCITY
      V=QU/H


! CALCULATION OF BED-SHEAR SRESS (N/m2)
       SHE=ROE*G*H*SEFF


! CALCULATION OF CRITICAL SHEAR STRESS (CSHE)
       KVIS=0.000001

      F=(D50/(KVIS))*(0.1*(SS-1)*G*D50)**0.5

      IF(F.EQ.500.OR.F.GT.500) A=0.06
     IF(F.LT.500.AND.F.GT.400.OR.F.EQ.400) A=0.057
     IF(F.LT.400.AND.F.GT.300.OR.F.EQ.300) A=0.055
      IF(F.LT.300.AND.F.GT.200.OR.F.EQ.200) A=0.050
      IF(F.LT.200.AND.F.GT.150.OR.F.EQ.150) A=0.045
      IF(F.LT.150.AND.F.GT.100.OR.F.EQ.100) A=0.040
      IF(F.LT.100.AND.F.GT.90.OR.F.EQ.90) A=0.039
      IF(F.LT.90.AND.F.GT.80.OR.F.EQ.80) A=0.038
      IF(F.LT.80.AND.F.GT.70.OR.F.EQ.70) A=0.037
      IF(F.LT.70.AND.F.GT.60.OR.F.EQ.60) A=0.035
      IF(F.LT.60.AND.F.GT.50.OR.F.EQ.50) A=0.035
      IF(F.LT.50.AND.F.GT.40.OR.F.EQ.40) A=0.033
      IF(F.LT.40.AND.F.GT.30.OR.F.EQ.30) A=0.032
      IF(F.LT.30.AND.F.GT.20.OR.F.EQ.20) A=0.031
      IF(F.LT.20.AND.F.GT.15.OR.F.EQ.15) A=0.031
      IF(F.LT.15.AND.F.GT.10.OR.F.EQ.10) A=0.035
      IF(F.LT.10.AND.F.GT.9.OR.F.EQ.9) A=0.035
      IF(F.LT.9.AND.F.GT.8.OR.F.EQ.8) A=0.036
      IF(F.LT.8.AND.F.GT.7.OR.F.EQ.7) A=0.038
      IF(F.LT.7.AND.F.GT.6.OR.F.EQ.6) A=0.036
         IF(F.LT.6.AND.F.GT.5.OR.F.EQ.5) A=0.045
      IF(F.LT.5.AND.F.GT.4.OR.F.EQ.4) A=0.050
      IF(F.LT.4.AND.F.GT.3.OR.F.EQ.3) A=0.058
      IF(F.LT.3.AND.F.GT.2.OR.F.EQ.2) A=0.070
      IF(F.LT.2.AND.F.GT.1.OR.F.EQ.1) A=0.082

      CSHE=(SS*ROE*G-ROE*G)*D50*A

    IF (SHE.LT.CSHE) THEN

    GO TO 1
      END IF

! CALCULATION OF USTAR
```

```
        USTAR=(SHE/ROE)**0.5

! CALCULATION OF AA VALUE
        AA=V*SHE*SF/(USTAR*(G*SS*ROE-G*ROE)*D50)

! DETERMINATION OF KT VALUES FROM THE GRAPH

        IF(D50.EQ.0.00019.AND.AA.EQ.0.0015.OR.AA.LT.0.0015) KT=0.01
        IF(D50.EQ.0.00019.AND.AA.GT.0.0015.AND.AA.LT.0.002) KT=0.013
        IF(D50.EQ.0.00019.AND.AA.EQ.0.002) KT=0.017
        IF(D50.EQ.0.00019.AND.AA.GT.0.002.AND.AA.LT.0.004) KT=0.027
        IF(D50.EQ.0.00019.AND.AA.EQ.0.004) KT=0.06
        IF(D50.EQ.0.00019.AND.AA.GT.0.004.AND.AA.LT.0.006) KT=0.085
        IF(D50.EQ.0.00019.AND.AA.EQ.0.006) KT=0.13
        IF(D50.EQ.0.00019.AND.AA.GT.0.006.AND.AA.LT.0.0067) KT=0.15
        IF(D50.EQ.0.00019.AND.AA.EQ.0.0067) KT=0.17
        IF(D50.EQ.0.00019.AND.AA.GT.0.0067.AND.AA.LT.0.01) KT=0.185
        IF(D50.EQ.0.00019.AND.AA.EQ.0.01) KT=0.2
        IF(D50.EQ.0.00019.AND.AA.GT.0.01.AND.AA.LT.0.04) KT=0.23
        IF(D50.EQ.0.00019.AND.AA.EQ.0.04) KT=0.4
        IF(D50.EQ.0.00019.AND.AA.GT.0.04.AND.AA.LT.0.1) KT=0.45
        IF(D50.EQ.0.00019.AND.AA.EQ.0.01) KT=0.6
        IF(D50.EQ.0.00019.AND.AA.GT.0.1.AND.AA.LT.0.2) KT=0.70
        IF(D50.EQ.0.00019.AND.AA.EQ.0.2) KT=0.8
        IF(D50.EQ.0.00019.AND.AA.GT.0.2.AND.AA.LT.0.4) KT=1
        IF(D50.EQ.0.00019.AND.AA.EQ.0.4) KT=1.2
        IF(D50.EQ.0.00019.AND.AA.GT.0.4.AND.AA.LT.0.6) KT=1.4
        IF(D50.EQ.0.00019.AND.AA.EQ.0.6) KT=1.5
        IF(D50.EQ.0.00019.AND.AA.GT.0.6) KT=1.5

        IF(D50.EQ.0.00033.AND.AA.EQ.0.001) KT=0.028
        IF(D50.EQ.0.00033.AND.AA.GT.0.001.AND.AA.LT.0.002) KT=0.01
        IF(D50.EQ.0.00033.AND.AA.EQ.0.002) KT=0.08
        IF(D50.EQ.0.00033.AND.AA.GT.0.002.AND.AA.LT.0.004) KT=0.15
        IF(D50.EQ.0.00033.AND.AA.EQ.0.004) KT=0.175
        IF(D50.EQ.0.00033.AND.AA.GT.0.004.AND.AA.LT.0.008) KT=0.25
        IF(D50.EQ.0.00033.AND.AA.EQ.0.008) KT=0.3
        IF(D50.EQ.0.00033.AND.AA.GT.0.008.AND.AA.LT.0.01) KT=0.415
        IF(D50.EQ.0.00033.AND.AA.EQ.0.01) KT=0.47
        IF(D50.EQ.0.00033.AND.AA.GT.0.01.AND.AA.LT.0.04) KT=0.65
        IF(D50.EQ.0.00033.AND.AA.EQ.0.04) KT=0.8
        IF(D50.EQ.0.00033.AND.AA.GT.0.04.AND.AA.LT.0.1) KT=0.85
        IF(D50.EQ.0.00033.AND.AA.EQ.0.1) KT=1
        IF(D50.EQ.0.00033.AND.AA.GT.0.1.AND.AA.LT.0.2) KT=1.2
        IF(D50.EQ.0.00033.AND.AA.EQ.0.2) KT=1.3
        IF(D50.EQ.0.00033.AND.AA.GT.0.2.AND.AA.LT.0.4) KT=1.5
        IF(D50.EQ.0.00033.AND.AA.EQ.0.4) KT=1.6
        IF(D50.EQ.0.00033.AND.AA.GT.0.4.AND.AA.LT.0.6) KT=1.65
        IF(D50.EQ.0.00033.AND.AA.EQ.0.06.OR.AA.GT.0.6) KT=1.65

        IF(D50.EQ.0.00036.AND.AA.EQ.0.001) KT=0.0135
        IF(D50.EQ.0.00036.AND.AA.GT.0.001.AND.AA.LT.0.002) KT=0.047
        IF(D50.EQ.0.00036.AND.AA.EQ.0.002) KT=0.065
        IF(D50.EQ.0.00036.AND.AA.GT.0.002.AND.AA.LT.0.004) KT=0.115
        IF(D50.EQ.0.00036.AND.AA.EQ.0.004) KT=0.17
        IF(D50.EQ.0.00036.AND.AA.GT.0.004.AND.AA.LT.0.008) KT=0.25
```

```
IF(D50.EQ.0.00036.AND.AA.EQ.0.008) KT=0.4
IF(D50.EQ.0.00036.AND.AA.GT.0.008.AND.AA.LT.0.01) KT=0.42
IF(D50.EQ.0.00036.AND.AA.EQ.0.01) KT=0.47
IF(D50.EQ.0.00036.AND.AA.GT.0.01.AND.AA.LT.0.04) KT=0.66
IF(D50.EQ.0.00036.AND.AA.EQ.0.04) KT=0.8
IF(D50.EQ.0.00036.AND.AA.GT.0.04.AND.AA.LT.0.08) KT=0.85
IF(D50.EQ.0.00036.AND.AA.EQ.0.08) KT=1
IF(D50.EQ.0.00036.AND.AA.GT.0.08.AND.AA.LT.0.1) KT=1.15
IF(D50.EQ.0.00036.AND.AA.EQ.0.1) KT=1.2
IF(D50.EQ.0.00036.AND.AA.GT.0.1.AND.AA.LT.0.2) KT=1.35
IF(D50.EQ.0.00036.AND.AA.EQ.0.2) KT=1.5
IF(D50.EQ.0.00036.AND.AA.GT.0.2.AND.AA.LT.0.4) KT=1.65
IF(D50.EQ.0.00036.AND.AA.EQ.0.4) KT=1.75
IF(D50.EQ.0.00036.AND.AA.GT.0.4.AND.AA.LT.0.6) KT=1.85
IF(D50.EQ.0.00036.AND.AA.EQ.0.6) KT=2
IF(D50.EQ.0.00036.AND.AA.GT.0.6) KT=2


IF(D50.EQ.0.00050.AND.AA.EQ.0.001) KT=0.12
IF(D50.EQ.0.00050.AND.AA.GT.0.001.AND.AA.LT.0.002) KT=0.15
IF(D50.EQ.0.00050.AND.AA.EQ.0.002) KT=0.17
IF(D50.EQ.0.00050.AND.AA.GT.0.002.AND.AA.LT.0.004) KT=0.2
IF(D50.EQ.0.00050.AND.AA.EQ.0.004) KT=0.22
IF(D50.EQ.0.00050.AND.AA.GT.0.004.AND.AA.LT.0.01) KT=0.27
IF(D50.EQ.0.00050.AND.AA.EQ.0.01) KT=0.38
IF(D50.EQ.0.00050.AND.AA.GT.0.01.AND.AA.LT.0.02) KT=0.48
IF(D50.EQ.0.00050.AND.AA.EQ.0.02) KT=0.56
IF(D50.EQ.0.00050.AND.AA.GT.0.02.AND.AA.LT.0.04) KT=0.65
IF(D50.EQ.0.00050.AND.AA.EQ.0.04) KT=0.8
IF(D50.EQ.0.00050.AND.AA.GT.0.04.AND.AA.LT.0.2) KT=0.7
IF(D50.EQ.0.00050.AND.AA.EQ.0.2) KT=0.6
IF(D50.EQ.0.00050.AND.AA.GT.0.2.AND.AA.LT.0.4) KT=0.57
IF(D50.EQ.0.00050.AND.AA.EQ.0.4) KT=0.54
IF(D50.EQ.0.00050.AND.AA.GT.0.4.AND.AA.LT.0.6) KT=0.5
IF(D50.EQ.0.00050.AND.AA.EQ.0.6) KT=0.48
IF(D50.EQ.0.00050.AND.AA.GT.0.6) KT=0.48


IF(D50.EQ.0.00052.AND.AA.EQ.0.001) KT=0.12
IF(D50.EQ.0.00052.AND.AA.GT.0.001.AND.AA.LT.0.002) KT=0.15
IF(D50.EQ.0.00052.AND.AA.EQ.0.002) KT=0.17
IF(D50.EQ.0.00052.AND.AA.GT.0.002.AND.AA.LT.0.004) KT=0.2
IF(D50.EQ.0.00052.AND.AA.EQ.0.004) KT=0.22
IF(D50.EQ.0.00052.AND.AA.GT.0.004.AND.AA.LT.0.01) KT=0.27
IF(D50.EQ.0.00052.AND.AA.EQ.0.01) KT=0.38
IF(D50.EQ.0.00052.AND.AA.GT.0.01.AND.AA.LT.0.02) KT=0.48
IF(D50.EQ.0.00052.AND.AA.EQ.0.02) KT=0.56
IF(D50.EQ.0.00052.AND.AA.GT.0.02.AND.AA.LT.0.04) KT=0.65
IF(D50.EQ.0.00052.AND.AA.EQ.0.04) KT=0.8
IF(D50.EQ.0.00052.AND.AA.GT.0.04.AND.AA.LT.0.2) KT=0.7
IF(D50.EQ.0.00052.AND.AA.EQ.0.2) KT=0.6
IF(D50.EQ.0.00052.AND.AA.GT.0.2.AND.AA.LT.0.4) KT=0.57
IF(D50.EQ.0.00052.AND.AA.EQ.0.4) KT=0.54
IF(D50.EQ.0.00052.AND.AA.GT.0.4.AND.AA.LT.0.6) KT=0.5
IF(D50.EQ.0.00052.AND.AA.EQ.0.6) KT=0.48
IF(D50.EQ.0.00052.AND.AA.GT.0.6) KT=0.48


IF(D50.EQ.0.00093.AND.AA.EQ.0.001) KT=0.1
```

```
        IF(D50.EQ.0.00093.AND.AA.GT.0.001.AND.AA.LT.0.006) KT=0.15
        IF(D50.EQ.0.00093.AND.AA.EQ.0.006) KT=0.2
        IF(D50.EQ.0.00093.AND.AA.GT.0.006.AND.AA.LT.0.01) KT=0.25
        IF(D50.EQ.0.00093.AND.AA.EQ.0.01) KT=0.3
        IF(D50.EQ.0.00093.AND.AA.GT.0.01.AND.AA.LT.0.04) KT=0.32
        IF(D50.EQ.0.00093.AND.AA.EQ.0.04) KT=0.4
        IF(D50.EQ.0.00093.AND.AA.GT.0.04.AND.AA.LT.0.08) KT=0.43
        IF(D50.EQ.0.00093.AND.AA.EQ.0.08) KT=0.5
        IF(D50.EQ.0.00093.AND.AA.GT.0.08.AND.AA.LT.0.15) KT=0.57
        IF(D50.EQ.0.00093.AND.AA.EQ.0.15) KT=0.6
        IF(D50.EQ.0.00093.AND.AA.GT.0.15.AND.AA.LT.0.4) KT=0.5
        IF(D50.EQ.0.00093.AND.AA.EQ.0.4) KT=0.4
        IF(D50.EQ.0.00093.AND.AA.GT.0.4.AND.AA.LT.0.6) KT=0.39
        IF(D50.EQ.0.00093.AND.AA.EQ.0.6) KT=0.37
        IF(D50.EQ.0.00093.AND.AA.GT.0.6) KT=0.37


!  CALCULATION OF SEDIMENT DISCHARGE
        QSU=KT*V*(SHE-CSHE)/(G*SS*ROE)
1       CONTINUE
        RETURN
        END


!16) COMPUTER PROGRAM CODE OF SUBROUTINE BAGNOLD

SUBROUTINE BAGNOLD(QU,H,ROE,G,SEFF,D50,SS,QSU)


!  SUBROUTINE SUBPROGRAM FOR CALCULATING THE TOTAL-LOAD TRANSPORT AT
!  A SECTION USING THE FORMULA OF BAGNOLD (1966)
!  LIST OF VARIABLES LOCALLY DEFINED FOR SUBROUTINE SUBPROGRAM
!  "BAGNOLD"
!  V = AVERAGE FLOW VELOCITY
!  SHE= BED SHEAR STRESS
!  W = FALL VELOCITY OF SEDIMENT PARTICLES
!  K = FIRST TERM OF RUBEYS'S FALL VELOCITY FORMULA
!  U = CONSTANT (=079) IN RUBEY'S FALL VELOCITY FORMULA
!  B = SECOND TERM OF RUBEYS'S FALL VELOCITY FORMULA
!  A = DIMENSIONLESS BED SHEAR STRESS
!  EB= EFFICIENCY COEFFICIENT
!  TALPHA= RATIO OF TANGENTIAL TO NORMAL FORCE
!  QBW =BED LOAD TRANSPORT RATE BY WEIGHT PER UNIT CHANNEL
!       WIDTH
!  DECLERATION OF VARIABLES
        REAL V,EB,TALPHA,SHE,KVIS,U,K,B,W
        DOUBLE PRECISION QBW,QSU,A
!  CALCULATION OF AVERAGE FLOW VELOCITY
        V=QU/H
!  CALCULATION OF BED-SHEAR SRESS (N/m2)
        SHE=ROE*G*H*SEFF

!  CALCULATION OF FALL VELOCITY, W (m/s)
        KVIS=0.000001

        K=(0.667+36*KVIS**2/(G*D50**3*(SS-1)))**0.5
        B=(36*KVIS**2/(G*D50**3*(SS-1)))**0.5

        IF (D50.EQ.0.001.OR.D50.GT.0.001) U=0.79
```

```
      IF (D50.LT.0.001) U=K-B

      W=U*(D50*G*(SS-1))**0.5

! DETERMINATION OF TALPHA VALUES
      A=SHE/((G*SS*ROE-G*ROE)*D50)

      IF (D50.EQ.0.00025) TALPHA=0.75
      IF (D50.EQ.0.0003) THEN
      IF (A.EQ.2) TALPHA=0.72
      IF (A.LT.2.AND.A.GT.1.5) TALPHA=0.73
      IF (A.EQ.1.5) TALPHA=0.74
      IF (A.LT.1.5.AND.A.GT.1) TALPHA=0.745
      IF (A.EQ.1.OR.A.LT.1) TALPHA=0.75
      END IF

      IF (D50.EQ.0.0004) THEN
      IF (A.EQ.2) TALPHA=0.68
      IF (A.LT.2.AND.A.GT.1.5) TALPHA=0.69
      IF (A.EQ.1.5) TALPHA=0.7
      IF (A.LT.1.5.AND.A.GT.1) TALPHA=0.72
      IF (A.EQ.1) TALPHA=0.73
      IF (A.LT.1.AND.A.GT.0.6) TALPHA=0.74
      IF (A.EQ.0.6.OR.A.LT.0.6) TALPHA=0.75
      END IF

      IF (D50.EQ.0.0005) THEN
      IF (A.EQ.2) TALPHA=0.57
      IF (A.LT.2.AND.A.GT.1.5) TALPHA=0.6
      IF (A.EQ.1.5) TALPHA=0.62
      IF (A.LT.1.5.AND.A.GT.1) TALPHA=0.64
      IF (A.EQ.1) TALPHA=0.66
      IF (A.LT.1.AND.A.GT.0.6) TALPHA=0.68
      IF (A.EQ.0.6) TALPHA=0.71
      IF (A.LT.0.6.AND.A.GT.0.3) TALPHA=0.73
      IF (A.EQ.0.3.OR.A.LT.0.3) TALPHA=0.74
      END IF

      IF (D50.EQ.0.0006) THEN
      IF (A.EQ.2) TALPHA=0.5
      IF (A.LT.2.AND.A.GT.1) TALPHA=0.55
      IF (A.EQ.1) TALPHA=0.58
      IF (A.LT.1.AND.A.GT.0.8) TALPHA=0.6
      IF (A.EQ.0.8) TALPHA=0.62
      IF (A.LT.0.8.AND.A.GT.0.6) TALPHA=0.63
      IF (A.EQ.0.6) TALPHA=0.65
      IF (A.LT.0.6.AND.A.GT.0.4) TALPHA=0.66
      IF (A.EQ.0.4) TALPHA=0.69
      IF (A.LT.0.4.AND.A.GT.0.3) TALPHA=0.70
      IF (A.EQ.0.3) TALPHA=0.72
      IF (A.LT.0.3.AND.A.GT.0.2) TALPHA=0.73
      IF (A.EQ.0.2) TALPHA=0.74
      IF (A.LT.0.2.AND.A.GT.0.1) TALPHA=0.745
      IF (A.EQ.0.1.OR.A.LT.0.1) TALPHA=0.75
      END IF
```

```
IF (D50.EQ.0.0007) THEN
 IF (A.EQ.2) TALPHA=0.45
 IF (A.LT.2.AND.A.GT.1) TALPHA=0.49
 IF (A.EQ.1) TALPHA=0.53
IF (A.LT.1.AND.A.GT.0.8) TALPHA=0.54
IF (A.EQ.0.8) TALPHA=0.55
IF (A.LT.0.8.AND.A.GT.0.6) TALPHA=0.57
IF (A.EQ.0.6) TALPHA=0.6
IF (A.LT.0.6.AND.A.GT.0.4) TALPHA=0.63
IF (A.EQ.0.4) TALPHA=0.65
IF (A.LT.0.4.AND.A.GT.0.3) TALPHA=0.66
IF (A.EQ.0.3) TALPHA=0.68
IF (A.LT.0.3.AND.A.GT.0.2) TALPHA=0.7
IF (A.EQ.0.2.OR.A.LT.0.2) TALPHA=0.71
 END IF

IF (D50.EQ.0.0008) THEN
 IF (A.EQ.2) TALPHA=0.41
 IF (A.LT.2.AND.A.GT.1) TALPHA=0.45
 IF (A.EQ.1) TALPHA=0.47
IF (A.LT.1.AND.A.GT.0.8) TALPHA=0.48
IF (A.EQ.0.8) TALPHA=0.5
IF (A.LT.0.8.AND.A.GT.0.6) TALPHA=0.53
IF (A.EQ.0.6) TALPHA=0.55
IF (A.LT.0.6.AND.A.GT.0.4) TALPHA=0.57
IF (A.EQ.0.4) TALPHA=0.6
IF (A.LT.0.4.AND.A.GT.0.3) TALPHA=0.62
IF (A.EQ.0.3) TALPHA=0.64
IF (A.LT.0.3.AND.A.GT.0.2) TALPHA=0.66
IF (A.EQ.0.2.OR.A.LT.0.2) TALPHA=0.68
 END IF

IF (D50.EQ.0.001) THEN
 IF (A.EQ.2) TALPHA=0.37
 IF (A.LT.2.AND.A.GT.1) TALPHA=0.38
 IF (A.EQ.1) TALPHA=0.42
IF (A.LT.1.AND.A.GT.0.6) TALPHA=0.44
IF (A.EQ.0.6) TALPHA=0.46
IF (A.LT.0.6.AND.A.GT.0.4) TALPHA=0.48
IF (A.EQ.0.4) TALPHA=0.5
IF (A.LT.0.4.AND.A.GT.0.3) TALPHA=0.53
IF (A.EQ.0.3) TALPHA=0.55
IF (A.LT.0.3.AND.A.GT.0.2) TALPHA=0.57
IF (A.EQ.0.2.OR.A.LT.0.2) TALPHA=0.6
 END IF

IF (D50.EQ.0.0012) THEN
 IF (A.EQ.2) TALPHA=0.37
 IF (A.LT.2.AND.A.GT.0.8) TALPHA=0.38
 IF (A.EQ.0.8) TALPHA=0.4
IF (A.LT.0.8.AND.A.GT.0.4) TALPHA=0.43
IF (A.EQ.0.4) TALPHA=0.45
IF (A.LT.0.4.AND.A.GT.0.3) TALPHA=0.47
IF (A.EQ.0.3) TALPHA=0.48
IF (A.LT.0.3.AND.A.GT.0.2) TALPHA=0.5
IF (A.EQ.0.2.OR.A.LT.0.2) TALPHA=0.5
```

```
      END IF

      IF (D50.GT.0.0015.OR.D50.EQ.0.0015) THEN
      IF (A.EQ.2) TALPHA=0.37
      IF (A.LT.2.AND.A.GT.0.6) TALPHA=0.37
      IF (A.EQ.0.6) TALPHA=0.38
      IF (A.LT.0.6.AND.A.GT.0.4) TALPHA=0.39
      IF (A.EQ.0.4) TALPHA=0.4
      IF (A.LT.0.4.AND.A.GT.0.2) TALPHA=0.43
      IF (A.EQ.0.2.OR.A.LT.0.2) TALPHA=0.44
      END IF

! eb VALUE IS ASSUMED AS 0.125
      EB=0.125

! CALCULATION OF QBW,QSW AND QSU VALUES
      QBW=(ROE*G*SHE*V*EB)/(TALPHA*(ROE*G*SS-ROE*G))
      QSW=0.01*ROE*G*SHE*V**2/(W*(ROE*G*SS-ROE*G))

      QSU=(QBW+QSW)/(G*ROE*SS)
      RETURN
      END

!17) COMPUTER PROGRAM CODE OF SUBROUTINE WUWANG

SUBROUTINE WUWANG(G,D50,SS,QU,H,QSU)
! V = AVERAGE FLOW VELOCITY
! W = FALL VELOCITY OF SEDIMENT PARTICLES
! A = FIRST TERM OF RUBEYS'S FALL VELOCITY FORMULA
! F = CONSTANT (=079) IN RUBEY'S FALL VELOCITY FORMULA
! B = SECOND TERM OF RUBEYS'S FALL VELOCITY FORMULA
! KVIS= KINEMATIC VISCOSITY OF WATER
! USP = UNIVERSAL STREAM POWER
! CT = SEDIMENT CONCENTRATION IN ppmC
! DECLERATION OF VARIABLES
      REAL V,W,A,B
       DOUBLE PRECISION CT,QSU,F
!CALCULATION OF FALL VELOCITY
      KVIS=0.000001

      A=(0.667+36*KVIS**2/(G*D50**3*(SS-1)))**0.5
      B=(36*KVIS**2/(G*D50**3*(SS-1)))**0.5

      IF (D50.EQ.0.001.OR.D50.GT.0.001) F=0.79
      IF (D50.LT.0.001) F=A-B

      W=F*(D50*G*(SS-1))**0.5

! CALCULATION OF AVERAGE FLOW VELOCITY
      V=QU/H

! CALCULATION OF UNIVERSAL STREAM POWER
      USP=V**3/((SS-1)*G*H*W*(LOG10(H/D50))**2)

!   CALCULATION OF SEDIMENT CONCENTRATION
      CT=(1430*(0.86+USP**0.5)*USP**1.5)/(0.016+USP)
```

```
!    CALCULATION OF SEDIMENT DISCHARGE
     QSU=QU*CT*0.000001


     RETURN
     END

!18) COMPUTER PROGRAM CODE OF SUBROUTINE BROWNLIE

SUBROUTINE BROWNLIE(QU,H,G,D50,SS,SF,QSU)
! CF=BROWNLIE'S COEFFICIENT FOR FIELD APPLICATIONS (=1.268)
! FG=THE GRAIN FROUDE NUMBER
! FGC= CRITICAL GRAIN FROUDE NUMBER
! C= SEDIMENT CONCENTRATION IN PARTS PER MILLION BY WEIGHT
! V= AVERAGE FLOW VELOCITY
! CSHE= CRITICAL SHEAR STRESS
! SDG=GEOMETRIC STANDARD DEVIATION OF SEDIMENT PARTICLE
!     SIZES (=1)
! Y = TEMPORARY VARIBLE USED IN THE METHOD
! RG= VARIABLE USED IN THE CALCULATION OF TEMPORARY
!     VARIABLE
!C DECLERATION OF VARIABLES
     REAL WIDTH,V,RG,KVIS,Y,SDG,CF
        DOUBLE PRECISION QSU,CSHE,FGC,FG,C


        KVIS=0.000001
!
! CALCULATION OF AVERAGE FLOW VELOCITY
     V=QU/H
!
! CALCULATION OF CRITICAL SHEAR STRESS
     RG=(G*D50**3)**0.5/KVIS
     Y=(1/((SS-1)**0.5*RG))**0.6
            CSHE=0.22*Y+0.06/(10**(7.7*Y))
            SDG=1
     FGC=4.596*CSHE**0.5293/(SF**0.1405*SDG**0.1606)
       FG=V/((SS-1)*G*D50)**0.5


     IF (FG.LT.FGC) THEN
            GOTO 1
     END IF
! CALCULATION OF SEDIMENT DISCHARGE
     CF=1.268
     C=7115*CF*(FG-FGC)**1.978*SF**0.6601*((D50*1000)/H)**0.3301
     QSU=C*QU/(10**6*SS)
     1    CONTINUE
        RETURN
     END

!19) COMPUTER PROGRAM CODE OF SUBROUTINE PARKER

SUBROUTINE PARKER(ROE,G,H,SEFF,SS,D50,QSU)
! SUBROUTINE SUBPROGRAM FOR CALCULATING THE BED-LOAD TRANSPORT AT A
! SECTION USING THE FORMULA OF PARKER (1982)
! SHE            BED SHEAR STRESS
! SHER           SHEAR PARAMETER (=0.0875)
```

```
! PHI              DIMENSIONLESS SHEAR STRESS PARAMETER
! USTAR            SHEAR VELOCITY
! W                DIMENSIONLESS BED-LOAD TRANSPORT NUMBER

! DECLERATION OF VARIABLES
     REAL SHER,SHE
     DOUBLE PRECISION QSU,W,PHI,USTAR
! CALCULATION OF BED-SHEAR SRESS (N/m2)
     SHE=ROE*G*H*SEFF
     SHER=0.0875
     PHI=SHE/((SS-1)*D50*SHER*G*ROE*SHER)
! CALCULATION OF SHEAR VELOCITY
     USTAR=(SHE/ROE)**0.5

     IF (PHI.EQ.0.95.OR.PHI.LT.0.95) THEN
   W=0.0025*EXP((14.2*(PHI-1)-9.28*(PHI-1)**2))
     END IF

     IF (PHI.GT.0.95.AND.PHI.LT.1.65) THEN
     W=0.0025*EXP((14.2*(PHI-1)-9.28*(PHI-1)**2))
     ENDIF

     IF (PHI.GT.1.65.OR.PHI.EQ.1.65) THEN
     W=11.2*(1-0.822/PHI)**4.5
     END IF

! CALCULATION OF SEDIMENT DISCHARGE
     QSU=W*USTAR*SHE/((SS-1)*ROE*G)

     RETURN
     END

!20) COMPUTER PROGRAM CODE OF SUBROUTINE EINSBROWN
SUBROUTINE EINSBROWN(G,D50,SS,QU,H,ROE,SEFF,QSU)

! SHE= BED SHEAR STRESS
! K = PARAMETER FROM RUBEY'S FORMULA FOR FALL VELOCITY
! KVIS = KINEMATIC VISCOSITY OF WATER
! PHI = INTENSITY OF BED-LOAD TRANSPORT
! KSI = FLOW INTENSITY
! RKSI= RECIPRAOCAL OF KSI
!
     REAL K,KVIS
     DOUBLE PRECISION SHE,PHI,KSI,RKSI,QSU

! CALCULATION OF BED-SHEAR SRESS (N/m2)
     SHE=ROE*G*H*SEFF
     KSI=(ROE*G*SS-ROE*G)*D50/SHE
     RKSI=1/KSI

! DETERMINATION OF PHI VALUES

     IF (RKSI.EQ.0.09) PHI=0.03
     IF (RKSI.GT.0.08.AND.RKSI.LT.0.09) PHI=0.025
     IF (RKSI.EQ.0.08) PHI=0.02
     IF (RKSI.GT.0.07.AND.RKSI.LT.0.08) PHI=0.015
```

```
      IF (RKSI.EQ.0.07) PHI=0.0095
      IF (RKSI.GT.0.06.AND.RKSI.LT.0.07) PHI=0.0080
       IF (RKSI.EQ.0.06) PHI=0.0065
      IF (RKSI.GT.0.05.AND.RKSI.LT.0.06) PHI=0.0045
      IF (RKSI.EQ.0.05) PHI=0.0025
      IF (RKSI.GT.0.04.AND.RKSI.LT.0.05) PHI=0.0010
      IF (RKSI.EQ.0.04.OR.RKSI.LT.0.04) PHI=0.0001

      IF (RKSI.GT.0.09) THEN
       PHI=40*RKSI**3
      END IF

! CALCULATION OF SEDIMENT DISCHARGE
      KVIS=0.000001
      K=(0.6667+36*KVIS**2/(G*D50**3*(SS-1)))**0.5-(36*KVIS**2/
       (G*D50**3*(SS-1)))**0.5
      QSU=PHI*K*(G*(SS-1)*D50**3)**0.5

      RETURN
      END

!21) COMPUTER PROGRAM CODE OF SUBROUTINE YALIN

SUBROUTINE YALIN(ROE,G,H,SEFF,D50,SS,QSU)

!C SUBROUTINE SUBPROGRAM FOR CALCULATING THE BED-LOAD TRANSPORT AT A
!C SECTION USING THE FORMULA OF YALIN (1963)
!C LIST OF VARIABLES LOCALLY DEFINED FOR SUBROUTINE SUBPROGRAM "YALIN"
!C VARIABLE NAME EXPLANATIONS
! QB = SEDIMENT DISCHARGE WEIGHT PER UNIT WIDTH
! USTAR=SHEAR VELOCITY
! SHE = BED SHEAR STRESS
! CSHE =CRITICAL BED SHEAR STRESS
! THETA=SHIELD'S PARAMETER DEPENDING ON BED SHEAR STRESS
! CTHETA= SHIELD'S PARAMETER DEPENDING ON CRICAL BED SHEAR
!     STRESS
! S = EQUATION PARAMETER
! AS= EQUATION PARAMETER
! F = VANONI'S (1977) PARAMETER TO DETERMINE CRITICAL SHEAR
!     STRESS
! A = DIMENSIONLESS SHIELDS PARAETER OF CORRESPONDING F
!
!

      REAL SHE,CSHE,Q,KVIS,WIDTH,F,A,USTAR
      DOUBLE PRECISION QSU,QB,S,AS,U,TERM1

      KVIS=0.000001
!
! CALCULATION OF BED-SHEAR SRESS (N/m2)
      SHE=ROE*G*H*SEFF
!
! CALCULATION OF CRITICAL SHEAR STRESS (CSHE)
      F=(D50/(KVIS))*(0.1*(SS-1)*G*D50)**0.5

      IF(F.EQ.500.OR.F.GT.500) A=0.06
      IF(F.LT.500.AND.F.GT.400.OR.F.EQ.400) A=0.057
```

```
       IF(F.LT.400.AND.F.GT.300.OR.F.EQ.300) A=0.055
       IF(F.LT.300.AND.F.GT.200.OR.F.EQ.200) A=0.050
       IF(F.LT.200.AND.F.GT.150.OR.F.EQ.150) A=0.045
       IF(F.LT.150.AND.F.GT.100.OR.F.EQ.100) A=0.040
       IF(F.LT.100.AND.F.GT.90.OR.F.EQ.90) A=0.039
       IF(F.LT.90.AND.F.GT.80.OR.F.EQ.80) A=0.038
       IF(F.LT.80.AND.F.GT.70.OR.F.EQ.70) A=0.037
       IF(F.LT.70.AND.F.GT.60.OR.F.EQ.60) A=0.035
       IF(F.LT.60.AND.F.GT.50.OR.F.EQ.50) A=0.035
       IF(F.LT.50.AND.F.GT.40.OR.F.EQ.40) A=0.033
       IF(F.LT.40.AND.F.GT.30.OR.F.EQ.30) A=0.032
       IF(F.LT.30.AND.F.GT.20.OR.F.EQ.20) A=0.031
       IF(F.LT.20.AND.F.GT.15.OR.F.EQ.15) A=0.031
       IF(F.LT.15.AND.F.GT.10.OR.F.EQ.10) A=0.035
       IF(F.LT.10.AND.F.GT.9.OR.F.EQ.9) A=0.035
       IF(F.LT.9.AND.F.GT.8.OR.F.EQ.8) A=0.036
       IF(F.LT.8.AND.F.GT.7.OR.F.EQ.7) A=0.038
       IF(F.LT.7.AND.F.GT.6.OR.F.EQ.6) A=0.036
      IF(F.LT.6.AND.F.GT.5.OR.F.EQ.5) A=0.045
       IF(F.LT.5.AND.F.GT.4.OR.F.EQ.4) A=0.050
       IF(F.LT.4.AND.F.GT.3.OR.F.EQ.3) A=0.058
       IF(F.LT.3.AND.F.GT.2.OR.F.EQ.2) A=0.070
       IF(F.LT.2.AND.F.GT.1.OR.F.EQ.1) A=0.082

       CSHE=(SS*ROE*G-ROE*G)*D50*A
! CALCULATION OF THETA AND CTHETA
       THETA=SHE/((ROE*G*SS-ROE*G)*D50)
       CTHETA=CSHE/((ROE*G*SS-ROE*G)*D50)
! CALCULATION OF s AND a VALUES
       S=(THETA-CTHETA)/CTHETA
       IF (S.LT.0.OR.S.EQ.0) THEN
       GOTO 1
       END IF

       AS=2.45*CTHETA**0.5*(1/SS)**0.4
       U=1+A*AS
       IF (U.LT.0.OR.U.EQ.0) THEN
       GOTO 1
       END IF

       TERM1=1-LOG(U)/(A*AS)

       IF (TERM1.LT.0) THEN
       GOTO 1
       END IF
!    CALCULATION OF USTAR
       USTAR=(SHE/ROE)**0.5
!    CALCULATION OF SEDIMENT DISCHARGE
       QSU=0.635*S*TERM1*D50*USTAR

1    CONTINUE

       RETURN
       END

!22) COMPUTER PROGRAM CODE OF SUBROUTINE ENGEFRED
```

```
SUBROUTINE ENGEFRED(ROE,G,H,SEFF,D50,SS,QSU)
! USTAR = SHEAR VELOCITY
! SHE = BED SHEAR STRESS
! CSHE = CRITICAL BED SHEAR STRESS
! THETA = SHIELD'S PARAMETER DEPENDING ON BED SHEAR STRESS
! CTHETA= SHIELD'S PARAMETER DEPENDING ON CRICAL BED SHEAR
!    STRESS
! F= VANONI'S (1977) PARAMETER TO DETERMINE CRITICAL SHEAR
!    STRESS
! A =DIMENSIONLESS SHIELDS PARAETER OF CORRESPONDING F
! FE=FUNCTION IN THE ENGELUND-FREDSOE FORMULA DEPENDING
!    ON CRITICAL SHIELD'S PARAMETER
! DECLERATION OF VARIABLES
      REAL G,SHE,CSHE,D50,H,ROE,SS,KVIS,QU,F,A
      DOUBLE PRECISION QSU,TERM1,FE

      KVIS=0.000001
! CALCULATION OF BED-SHEAR SRESS (N/m2)
      SHE=ROE*G*H*SEFF

! CALCULATION OF CRITICAL SHEAR STRESS (CSHE)

      F=(D50/(KVIS))*(0.1*(SS-1)*G*D50)**0.5

      IF(F.EQ.500.OR.F.GT.500) A=0.06
      IF(F.LT.500.AND.F.GT.400.OR.F.EQ.400) A=0.057
      IF(F.LT.400.AND.F.GT.300.OR.F.EQ.300) A=0.055
      IF(F.LT.300.AND.F.GT.200.OR.F.EQ.200) A=0.050
      IF(F.LT.200.AND.F.GT.150.OR.F.EQ.150) A=0.045
      IF(F.LT.150.AND.F.GT.100.OR.F.EQ.100) A=0.040
      IF(F.LT.100.AND.F.GT.90.OR.F.EQ.90) A=0.039
      IF(F.LT.90.AND.F.GT.80.OR.F.EQ.80) A=0.038
      IF(F.LT.80.AND.F.GT.70.OR.F.EQ.70) A=0.037
      IF(F.LT.70.AND.F.GT.60.OR.F.EQ.60) A=0.035
      IF(F.LT.60.AND.F.GT.50.OR.F.EQ.50) A=0.035
      IF(F.LT.50.AND.F.GT.40.OR.F.EQ.40) A=0.033
      IF(F.LT.40.AND.F.GT.30.OR.F.EQ.30) A=0.032
      IF(F.LT.30.AND.F.GT.20.OR.F.EQ.20) A=0.031
      IF(F.LT.20.AND.F.GT.15.OR.F.EQ.15) A=0.031
      IF(F.LT.15.AND.F.GT.10.OR.F.EQ.10) A=0.035
      IF(F.LT.10.AND.F.GT.9.OR.F.EQ.9) A=0.035
      IF(F.LT.9.AND.F.GT.8.OR.F.EQ.8) A=0.036
      IF(F.LT.8.AND.F.GT.7.OR.F.EQ.7) A=0.038
      IF(F.LT.7.AND.F.GT.6.OR.F.EQ.6) A=0.036
      IF(F.LT.6.AND.F.GT.5.OR.F.EQ.5) A=0.045
      IF(F.LT.5.AND.F.GT.4.OR.F.EQ.4) A=0.050
      IF(F.LT.4.AND.F.GT.3.OR.F.EQ.3) A=0.058
      IF(F.LT.3.AND.F.GT.2.OR.F.EQ.2) A=0.070
      IF(F.LT.2.AND.F.GT.1.OR.F.EQ.1) A=0.082

      CSHE=(SS*ROE*G-ROE*G)*D50*A
! CALCULATION OF THETA AND CTHETA
      THETA=SHE/((ROE*G*SS-ROE*G)*D50)
      CTHETA=CSHE/((ROE*G*SS-ROE*G)*D50)
```

```
      IF (CTHETA.GT.THETA) THEN
       GOTO 1
       END IF

       TERM1=THETA**0.5-0.7*CTHETA**0.5
       IF (TERM1.LT.0) THEN
       GOTO 1
       END IF
! CALCULATION OF SEDIMENT DISCHARGE
       FE=11.6*(THETA-CTHETA)*(TERM1)
       QSU=FE*((SS-1)*G*D50**3)**0.5


1      CONTINUE

       RETURN
       END


!23) COMPUTER PROGRAM CODE OF SUBROUTINE VANRIJN

SUBROUTINE VANRIJN(QU,H,D50,SS,G,ROE,QSU)
! F =VANONI'S (1977) PARAMETER TO DETERMINE CRITICAL SHEAR
!     STRESS
! A = DIMENSIONLESS SHIELDS PARAMETER OF CORRESPONDING F
! KVIS = KINEMATIC VISCOSITY OF WATER
! V = AVERAGE FLOW VELOCITY
! CUSTAR= BED SHEAR VELOCITY
!GRUSTAR= EFFECTIVE BED SHEAR VELOCITY RELATED TO GRAIN
!     ROUGHNESS
! CSHE = CRITICAL BED SHEAR STRESS
! T = DIMENSIONLESS EXCESS BED SHEAR STRESS
! DSTAR = DIMENSIONLESS PARICLE DIAMETERC

       REAL KVIS,F,A,CUSTAR,WIDTH,Q,V,GRUSTAR,T
       DOUBLE PRECISION QSU,CSHE,DSTAR
       KVIS=0.000001
! CALCULATION OF AVERAGE FLOW VELOCITY
       V=QU/H
! CALCULATION OF CRITICAL SHEAR STRESS (CSHE)
       F=(D50/(KVIS))*(0.1*(SS-1)*G*D50)**0.5

       IF(F.EQ.500.OR.F.GT.500) A=0.06
       IF(F.LT.500.AND.F.GT.400.OR.F.EQ.400) A=0.057
       IF(F.LT.400.AND.F.GT.300.OR.F.EQ.300) A=0.055
       IF(F.LT.300.AND.F.GT.200.OR.F.EQ.200) A=0.050
       IF(F.LT.200.AND.F.GT.150.OR.F.EQ.150) A=0.045
       IF(F.LT.150.AND.F.GT.100.OR.F.EQ.100) A=0.040
       IF(F.LT.100.AND.F.GT.90.OR.F.EQ.90) A=0.039
       IF(F.LT.90.AND.F.GT.80.OR.F.EQ.80) A=0.038
       IF(F.LT.80.AND.F.GT.70.OR.F.EQ.70) A=0.037
       IF(F.LT.70.AND.F.GT.60.OR.F.EQ.60) A=0.035
       IF(F.LT.60.AND.F.GT.50.OR.F.EQ.50) A=0.035
       IF(F.LT.50.AND.F.GT.40.OR.F.EQ.40) A=0.033
       IF(F.LT.40.AND.F.GT.30.OR.F.EQ.30) A=0.032
       IF(F.LT.30.AND.F.GT.20.OR.F.EQ.20) A=0.031
       IF(F.LT.20.AND.F.GT.15.OR.F.EQ.15) A=0.031
       IF(F.LT.15.AND.F.GT.10.OR.F.EQ.10) A=0.035
```

```
      IF(F.LT.10.AND.F.GT.9.OR.F.EQ.9) A=0.035
      IF(F.LT.9.AND.F.GT.8.OR.F.EQ.8) A=0.036
      IF(F.LT.8.AND.F.GT.7.OR.F.EQ.7) A=0.038
      IF(F.LT.7.AND.F.GT.6.OR.F.EQ.6) A=0.036
     IF(F.LT.6.AND.F.GT.5.OR.F.EQ.5) A=0.045
      IF(F.LT.5.AND.F.GT.4.OR.F.EQ.4) A=0.050
      IF(F.LT.4.AND.F.GT.3.OR.F.EQ.3) A=0.058
      IF(F.LT.3.AND.F.GT.2.OR.F.EQ.2) A=0.070
      IF(F.LT.2.AND.F.GT.1.OR.F.EQ.1) A=0.082


      CSHE=(SS*ROE*G-ROE*G)*D50*A
! CALCULATION OF CRITICAL BED SHEAR VELOCITY
      CUSTAR=(CSHE/ROE)**0.5
! CALCULATION OF EFFECTIVE BED SHEAR VELOCITY DUE TO GRAIN ROUGHNESS

      GRUSTAR=V/(5.75*LOG10(H/D50)+6.25)

      IF (GRUSTAR.LT.CUSTAR) THEN
      GOTO 1
      END IF
! CALCULATION OF T
      T=(GRUSTAR/CUSTAR)**2-1
! CALCULATION OF DSTAR
      DSTAR=D50*((SS-1)*G/KVIS**2)**0.3333
! CALCULATION OF SEDIMENT DISCHARGE
      QSU=0.053*((SS-1)*G)**0.5*D50**1.5*T**2.1/DSTAR**0.3


1     CONTINUE
      RETURN
      END

!24) COMPUTER PROGRAM CODE OF SUBROUTINE DOU

SUBROUTINE DOU(H,D50,SS,G,QU,SF,QSU)
! SUBROUTINE SUBPROGRAM FOR CALCULATING THE BED-LOAD TRANSPORT AT A
!
! VK= INCIPIENT FLOW VELOCITY
! M = COEFFICIENT FOR THE CALCULATION OF VK
! EPSILONK = COEFFICIENT FOR THE CALCULATION OF VK
! DELTA = COEFFICIENT FOR THE CALCULATION OF VK
! W = FALL VELOCITY OF SEDIMENT PARTICLES
! A = FIRST TERM OF RUBEYS'S FALL VELOCITY FORMULA
! F = CONSTANT (=079) IN RUBEY'S FALL VELOCITY FORMULA
! B = SECOND TERM OF RUBEYS'S FALL VELOCITY FORMULA
! KVIS = KINEMATIC VISCOSITY OF WATER
! CO = DIMENSIONLESS CHEZY COEFFICIENT
! KO = COEFFICIENT FOR COMPUTING SEDIMENT DISCHARGE
! V = AVERAGE FLOW VELOCITY
!
      REAL M,H,D50,SS,G,EPSILONK,DELTA,ROS,VK,V,QU,ROE
      REAL KO,CO,A,B,F,KVIS,W
      DOUBLE PRECISION QSU
!
! CALCULATION OF INITIAL VELOCITY
      M=0.32
      EPSILONK=0.00000256
```

```
      DELTA=0.00000021

      VK=M*LOG(11*H/D50)*((SS-1)*G*D50+0.19*(EPSILONK+G*H*DELTA)/D50)**0.5

! CALCULATION OF AVERAGE VELOCITY, V
      V=QU/H
      IF (V.LT.VK) THEN
      GOTO 1
      END IF
! CALCULATION OF FALL VELOCITY, W
      KVIS=0.000001
      A=(0.667+36*KVIS**2/(G*D50**3*(SS-1)))**0.5
      B=(36*KVIS**2/(G*D50**3*(SS-1)))**0.5
      IF (D50.EQ.0.001.OR.D50.GT.0.001) F=0.79
      IF (D50.LT.0.001) F=A-B
      W=F*(D50*G*(SS-1))**0.5
! CALCULATION OF DIMENSIONLESS CHEZY COEFFICIENT,CO
      CO=V/(H*SF*G)**0.5
! CALCULATION OF BEDLOAD D?SCHARGE,QSU
      KO=0.01
      QSU=(KO/CO**2)/(SS-1)*(V-VK)*V**3/(G*W)


1     CONTINUE

      RETURN
      END


!25) COMPUTER PROGRAM CODE OF SUBROUTINE KARKEN

SUBROUTINE KARKEN(SEFF,SS,G,D50,QU,H,ROE,QSU)
! KVIS= VISCOSITY OF WATER
! VA = FLOW VELOCITY CALCULATED IN THE APPROACH'S
!     CALCULATION PROCEDURE
! CSHE = CRITICAL BED SHEAR STRESS
! USTAR= BED SHEAR VELOCITY
! CUSTAR =CRITICAL BED SHEAR VELOCITY
! F = VANONI'S (1977) PARAMETER TO DETERMINE CRITICAL SHEAR
! STRESS
! A= DIMENSIONLESS SHIELDS PARAMETER OF CORRESPONDING F
! TERM1=RIGHT HAND SIDE OF THE SEDIMENT DISCHARGE EQUATION
!
!
!DECLERATION OF VARIABLES
      REAL V,VA,F,KVIS,A
      DOUBLE PRECISION CSHE, USTAR, CUSTAR, QSU,TERM1
      KVIS=0.000001
! CALCULATION OF FLOW VELOCITY, VA
      VA=((SS-1)*G*D50)**0.5*2.822*(QU/((SS-1)*G*D50**3)**0.5)**0.376
          *SEFF**0.310
! CALCULATION OF BED SHEAR VELOCITY, USTAR
      USTAR=(G*H*SEFF)**0.5

! CALCULATION OF CRITICAL BED SHEAR STRESS, CSHE
      F=(D50/(KVIS))*(0.1*(SS-1)*G*D50)**0.5

      IF(F.EQ.500.OR.F.GT.500) A=0.06
```

```
      IF(F.LT.500.AND.F.GT.400.OR.F.EQ.400) A=0.057
      IF(F.LT.400.AND.F.GT.300.OR.F.EQ.300) A=0.055
       IF(F.LT.300.AND.F.GT.200.OR.F.EQ.200) A=0.050
       IF(F.LT.200.AND.F.GT.150.OR.F.EQ.150) A=0.045
       IF(F.LT.150.AND.F.GT.100.OR.F.EQ.100) A=0.040
       IF(F.LT.100.AND.F.GT.90.OR.F.EQ.90) A=0.039
       IF(F.LT.90.AND.F.GT.80.OR.F.EQ.80) A=0.038
       IF(F.LT.80.AND.F.GT.70.OR.F.EQ.70) A=0.037
       IF(F.LT.70.AND.F.GT.60.OR.F.EQ.60) A=0.035
       IF(F.LT.60.AND.F.GT.50.OR.F.EQ.50) A=0.035
       IF(F.LT.50.AND.F.GT.40.OR.F.EQ.40) A=0.033
       IF(F.LT.40.AND.F.GT.30.OR.F.EQ.30) A=0.032
       IF(F.LT.30.AND.F.GT.20.OR.F.EQ.20) A=0.031
       IF(F.LT.20.AND.F.GT.15.OR.F.EQ.15) A=0.031
       IF(F.LT.15.AND.F.GT.10.OR.F.EQ.10) A=0.035
       IF(F.LT.10.AND.F.GT.9.OR.F.EQ.9) A=0.035
       IF(F.LT.9.AND.F.GT.8.OR.F.EQ.8) A=0.036
       IF(F.LT.8.AND.F.GT.7.OR.F.EQ.7) A=0.038
       IF(F.LT.7.AND.F.GT.6.OR.F.EQ.6) A=0.036
      IF(F.LT.6.AND.F.GT.5.OR.F.EQ.5) A=0.045
       IF(F.LT.5.AND.F.GT.4.OR.F.EQ.4) A=0.050
       IF(F.LT.4.AND.F.GT.3.OR.F.EQ.3) A=0.058
       IF(F.LT.3.AND.F.GT.2.OR.F.EQ.2) A=0.070
       IF(F.LT.2.AND.F.GT.1.OR.F.EQ.1) A=0.082

      CSHE=(SS*ROE*G-ROE*G)*D50*A
! CALCULATION OF CRITICAL BED SHEAR VELOCITY, CUSTAR
      CUSTAR=(CSHE/ROE)**0.5

      IF (USTAR.EQ.CUSTAR.OR.USTAR.LT.CUSTAR) THEN
      GOTO 1
      END IF
! CALCULATION OF SEDIMENT DISCHARGE, QSU
      TERM1=-2.279+2.972*LOG10(VA/((SS-1)*G*D50)**0.5)+1.060*LOG10(VA/((&
      SS-1)*G*D50)**0.5)*LOG10((USTAR-CUSTAR)/((SS-1)*G*D50)**0.5)+0.299 &
      *LOG10(H/D50)*LOG10((USTAR-CUSTAR)/((SS-1)*G*D50)**0.5)

          QSU=(0.1**ABS(TERM1))*((SS-1)*G*D50**3)**0.5

1     CONTINUE
      RETURN
      END

!26) COMPUTER PROGRAM CODE OF SUBROUTINE BISHOP
SUBROUTINE BISHOP(SEFF,SS,D50,H,G,QSU)
!
! PSIP= SHEAR INTENSITY FACTOR
! TPHI= INTENSITY OF TOTAL LOAD TRANSPORT PARAMETER
!
! DECLERATION OF VARIABLES
      REAL TPHI
      DOUBLE PRECISION QSU, PSIP

! CALCULATION OF SHEAR INTENSITY FACTOR, PSIP
      PSIP=SS*D50/(H*SEFF)
```

```
! DETERMINATION OF INTENSITY OF TOTAL LOAD TRANSPORT PARAMETER, TPHI
      IF (PSIP.GT.30) THEN
       GOTO 1
       END IF
      IF (PSIP.EQ.30) TPHI=0.0001
       IF (PSIP.LT.30.AND.PSIP.GT.20) TPHI=0.0005
       IF (PSIP.EQ.20) TPHI=0.0010
       IF (PSIP.LT.20.AND.PSIP.GT.15) TPHI=0.0080
      IF (PSIP.EQ.15) TPHI=0.0090
       IF (PSIP.LT.15.AND.PSIP.GT.10) TPHI=0.0200
       IF (PSIP.EQ.10) TPHI=0.0800
       IF (PSIP.LT.10.AND.PSIP.GT.7) TPHI=0.3000
       IF (PSIP.EQ.7) TPHI=0.7000
       IF (PSIP.LT.7.AND.PSIP.GT.5) TPHI=0.8500
       IF (PSIP.EQ.5) TPHI=1.1
       IF (PSIP.LT.5.AND.PSIP.GT.4) TPHI=2.0
       IF (PSIP.EQ.4) TPHI=3.0
       IF (PSIP.LT.4.AND.PSIP.GT.3) TPHI=4.0
       IF (PSIP.EQ.3) TPHI=8.0
       IF (PSIP.LT.3.AND.PSIP.GT.2) TPHI=8.5
       IF (PSIP.EQ.2) TPHI=9.0
       IF (PSIP.LT.2.AND.PSIP.GT.1) TPHI=10
       IF (PSIP.EQ.1) TPHI=70
       IF (PSIP.LT.1.AND.PSIP.GT.0.5) TPHI=150
      IF (PSIP.LT.0.5) THEN
       GOTO 1
       END IF


! CALCULATION OF SEDIMENT DISCHARGE, QSU
      QSU=TPHI*((SS-1)*G*D50**3)**0.5

1     CONTINUE
      RETURN
      END


!27) COMPUTER PROGRAM CODE OF SUBROUTINE WILCR

SUBROUTINE WILCR(G,ROE,H,SEFF,SS,D50,QSU)

! SHE= BED SHEAR STRESS
! SP = SHIELDS STRESS PARAMETER
! CSP= CRITICAL SHEAR STRESS PARAMETER
! KSI= RATIO SHIELD STRESS PARAMETER TO CRITICAL SHEAR STRESS
!     PARAMETER (SP/CSP)
! FKSI=VALUE OF THE FUNCTION DEPENDING ON CORREPONSING KSI
! DECLERATION OF VARIABLES

      REAL SHE,SP,KSI,FKSI
      DOUBLE PRECISION CSP,QSU


! CALCULATION OF BED SHEAR STRESS
     SHE=G*ROE*H*SEFF
! CALCULATION OF SHIELD STRESS PARAMETER, SP
     SP=SHE/(ROE*G*(SS-1)*D50)


! CALCULATION OF CRITICAL SHEAR STRESS PARAMETER, CSP
```

```
      CSP=0.021+0.015/EXP(20.0)
      KSI=SP/CSP
! DETERMINATION OF THE FUNCTION VALUES
      IF (KSI.GT.1.35.OR.KSI.EQ.1.35) THEN
      FKSI=(1-0.894/KSI**0.5)**4.5
      END IF
      IF (KSI.LT.1.35) THEN
      FKSI=0.000143*KSI**7.5
      END IF
!
! CALCULATION OF SEDIMENT DISCHARGE

      QSU=(14*FKSI*(SHE/ROE)**1.5)/(G*(SS-1))
      RETURN
      END


!28) COMPUTER PROGRAM CODE OF SUBROUTINE EINS50

SUBROUTINE EINS50 (G,H,SEFF,D50,ROS,ROE,SS,QSU)
! SUBROUTINE SUBPROGRAM FOR CALCULATING THE BED-LOAD TRANSPORT AT A
! SECTION USING THE FORMULA OF EINSTEIN (1950)

! LIST OF VARIABLES LOCALLY DEFINED FOR SUBROUTINE SUBPROGRAM "EINS50"
!
! VARIABLE NAME EXPLANATIONS
! USTARP= SHEAR VELOCITY DUE TO GRAIN ROUGHNESS
! KVIS = KINEMATIC VISCOSITY OF WATER
! DELTA = THICKNESS OF LAMINAR SUBLAYER
! CDELTA= APPERANT ROUGHNESS DIAMETER
! KS = ROUGHNESS DIAMETER
! XI = CORRECTION FACTOR
! PSI = DIMENSIONLESS BED-LOAD INTENSITY FACTOR
! PHIS = FLOW INTENSITY PARAMETER
! PSIS = CORRECTED DIMENSIONLESS BED-LOAD INTENSITY FACTOR
! BETA = LOGARITHMIC FUNCTION
! BETAX = LOGARITHMIC FUNCTION
! CX = REPRESENTATIVE GRAIN DIAMETER
! QB = BED-LOAD DISCHARGE BY WEIGHT PER UNIT CHANNEL WIDTH
!
!DECLARATION OF VARIABLES
      REAL PE,USTARP,KVIS,DELTA,KS,X5,XI
      REAL PSI,BETA,BETAX,CDELTA,PSIS,W,K,B,U,CX
      DOUBLE PRECISION QSU,QB

      KVIS=0.000001
! CALCULATION OF USTARP
      USTARP=(G*H*SEFF)**0.5

      DELTA=11.6*KVIS/USTARP
! CALCULATION OF CORRECTION FACTOR, X
      KS=D50
      XI=KS/DELTA
      CALL FIG1 (XI,X5)
! CALCULATION OF PSI
      PSI=(ROS-ROE)*D50/(ROE*H*SEFF)
      BETA=LOG(10.6)
```

```
      CDELTA=D50/X5
      CX=D50
      IF ((CDELTA/DELTA).GT.1.8) CX=0.77*CDELTA
      IF ((CDELTA/DELTA).LT.1.8) CX=1.39*DELTA
      BETAX=LOG(10.6*D50/CDELTA)
      PSIS=(BETA/BETAX)**2*PSI
!
! DETERMINATION OF PHIS VALUE
      CALL FIG2 (PSIS,PHIS)
! CALCULATION OF BED-LOAD
      QB=PHIS*((ROS-ROE)*G*D50**3/ROE)**0.5
      QSU=QB
      END


      SUBROUTINE FIG1 (XI,X5)
! THIS SUBROUTINE APPROXIMATES CORRECTION FACTOR, X
!    X=F(KS/DELTA)
      DIMENSION FX(8),FA(8),FB(8)
      DATA FX /0.5,0.65,0.9,1.15,1.4,3.2,5.0,8.4/
      DATA FA /1.9,1.75,1.62,1.61,1.63,1.72,1.42,1.25/
      DATA FB /1.72,1.23,0.57,0.0,-0.47,-1.11,-0.52,-0.27/
      N=0
      X5=0.4
      IF (XI.LT.0.135) RETURN
      IF (XI.LT.8.4) GOTO 100
      X5=1.0
      RETURN
100   N=N+1
      IF (XI.GT.FX(N)) GOTO 100
      X5=FB(N)*ALOG10(XI)+FA(N)
      RETURN
      END


      SUBROUTINE FIG2(PSIS,PHIS)
!
! THIS SUBROUTINE APPROXIMATES EINSTEIN PHISTAR (PHIS) VALUE
!    PHI=F(PSI)

      DIMENSION FX(7),FA(7),FB(7)
      DATA FX /0.77,2.12,4.1,6.1,11.0,16.7,22.5/
      DATA FA /7.56,5.35,4.1,4.1,4.6,5.66,9.28/
      DATA FB /1.01,1.19,1.67,2.3,3.23,4.26,7.81/
      N=0
      IF (PSIS.LT.22.5) GOTO 100
      PHIS=(13.1/PSIS)**12.66
      RETURN
100   N=N+1
      IF (PSIS.GT.FX(N)) GOTO 100
      PHIS=(FA(N)/PSIS)**FB(N)
      RETURN
      END


!29) COMPUTER PROGRAM CODE OF SUBROUTINE COLBY

SUBROUTINE COLBY(QU,H,D50,QSU,G,ROS)
```

```
!VC= CRITICAL FLOW VELOCITY
!V= AVERAGE FLOW VELOCITY
!TEMP= WATER TEMPERATURE
!

! DECLERATION OF VARIABLES

      REAL VC,V,DIFF,B,X,TEMP,F1,F2,CY,CF,AF
      DOUBLE PRECISION QSU
      DIMENSION CY(7,7),CF(5)
      DATA CF/0.64,1,1,0.88,0.2/
      DATA (CY(1,K),K=1,7) /0.1,0.2,0.3,0.4,0.8,0,0/
      DATA (CY(2,K),K=1,7) /0.61,0.48,0.3,0.3,0.3,0,0/
      DATA (CY(3,K),K=1,7) /1.453,1.329,1.4,1.26,1.099,0,0/
      DATA (CY(4,K),K=1,7) /0.01,5,10,15.6,20,30,40/
      DATA (CY(5,K),K=1,7) /0.1057,0.0845,0.0469,0,-0.0227,-0.0654,-0.1155/
      DATA (CY(6,K),K=1,7) /0.0735,0.0166,0.0014,0,-0.0164,-0.061,-0.0763/
      DATA (CY(7,K),K=1,7) /0.0118,0.0202,0.0135,0,0,0,0/
      TEMP=20.0

      V=QU/H

      VC=0.4673*H**0.1*D50**0.333
      DIFF=V-VC
      B=2.5
      IF (DIFF.GE.1.0) B=1.453*D50**(-0.138)
      H=H/0.3048
      X=LOG10(H)
      N=0
20    N=N+1
      IF (TEMP.GT.CY(4,N)) GOTO 20
      F1=CY(5,N-1)+CY(6,N-1)*X+CY(7,N-1)*X*X
      F2=CY(5,N)+CY(6,N)*X+CY(7,N)*X*X
      AF=F1+(F2-F1)*(LOG10(TEMP)-LOG10(CY(4,N-1)))/(LOG10(CY(4,N))
         -LOG10(CY(4,N-1)))
      AF=10**AF
      N=0
      D50=D50*1000
30    N=N+1
      IF (D50.GT.CY(1,N)) GOTO 30
      A=CY(3,N-1)*H**(CY(2,N-1))
      F1=A*DIFF**B*(1+(AF-1)*CF(N-1))*0.672
      A=CY(3,N)*H**(CY(2,N))
      F2=A*DIFF**B*(1+(AF-1)*CF(N))*0.672
      QSU=LOG10(F1)+(LOG10(F2)-LOG10(F1))*(LOG10(D50)-LOG10(CY(1,N-1)))/
         (LOG10(CY(1,N))-LOG10(CY(1,N-1)))
      QSU=10**QSU
      QSU=QSU*4.4487/(0.3048*G*ROS)
      PRINT*,"QSU=",QSU,"m3/s/m"
      D50=D50*0.001
      H=H*0.3048
      RETURN
      END

!30) COMPUTER PROGRAM CODE OF SUBROUTINE YANGG
```

```
SUBROUTINE YANGG (G,D50,SS,QU,H,ROE,SEFF,QSU)

!
! KVIS = KINEMATIC VISCOSITY OF WATER
! W    = FALL VELOCITY OF SEDIMENT PARTICLES
! K    = FIRST TERM OF RUBEYS'S FALL VELOCITY FORMULA
! U    = CONSTANT (=079) IN RUBEY'S FALL VELOCITY FORMULA
! B    = SECOND TERM OF RUBEYS'S FALL VELOCITY FORMULA
! VCR  = CRITICAL VELOCITY
! SHE  = BED SHEAR STRESS
! USTAR = SHEAR VELOCITY
! RATIO1= VCR/W
! SRE = SHEAR REYNOLDS NUMBER DEFINED AS USTAR*D50/KVIS
! TERM1 = FIRST TERM USED IN THE CALCULATION OF SEDIMENT
!            DISCHARGE
! TERM2 = SECOND TERM USED IN THE CALCULATION OF SEDIMENT
!     DISCHARGE
! AX = CHECK PARAMETER FOR SEDIMENT DISCHARGE CALCULATION
! MFW = MASS FLOW RATE
! CTS = TOTAL SAND CONCENTRATION IN ppm by Weight
!
! DECLERATION OF VARIABLES

    REAL SHE,USTAR,RATIO1,V,MFW
    DOUBLE PRECISION KVIS,B,U,K,AX
    DOUBLE PRECISION QSU,SRE,TERM1,TERM2,CTS
!
! CALCULATION OF FALL VELOCITY

    KVIS=0.000001

    K=(0.667+36*KVIS**2/(G*D50**3*(SS-1)))**0.5
    B=(36*KVIS**2/(G*D50**3*(SS-1)))**0.5

    IF (D50.EQ.0.001.OR.D50.GT.0.001) U=0.79
    IF (D50.LT.0.001) U=K-B

    W=U*(D50*G*(SS-1))**0.5
!
! CALCULATION OF BED-SHEAR STRESS (N/m2)
    V=QU/H
      SHE=ROE*G*H*SEFF

! CALCULATION OF SHEAR VELOCITY, USTAR
    USTAR=(SHE/ROE)**0.5
!
! CALCULATION OF SEDIMENT DISCHARGE
    SRE=USTAR*D50/KVIS

    IF (SRE.GT.1.2.AND.SRE.LT.70.OR.SRE.EQ.1.2) THEN
    RATIO1=2.5/(LOG10(SRE)-0.06)+0.66
    END IF

    IF (SRE.GT.70.OR.SRE.EQ.70) THEN
    RATIO1=2.05
    END IF
```

```
        IF (SRE.LT.1.2) THEN
        GO TO 1
        END IF

    TERM1=6.681-0.633*LOG10(W*D50/KVIS)-4.816*LOG10(USTAR/W)

    AX=V*SEFF/W-RATIO1*SEFF
    IF (AX.LT.0.OR.AX.EQ.0) THEN
    GOTO 1
    END IF

    TERM2=(2.784-0.305*LOG10(W*D50/KVIS)-0.282*LOG10(USTAR/W))*LOG10(V*SEFF/
          W-RATIO1*SEFF)

    CTS=TERM1+TERM2
    CTS=10**CTS

    MFW=QU*ROE
    QSU=MFW*CTS/(1000000*SS*ROE)

1       CONTINUE
        RETURN
        END

!31) COMPUTER PROGRAM CODE OF SUBROUTINE YANGLIM

SUBROUTINE YANGLIM (ROE,G,H,SEFF,D50,SS,QSU,ROS)

!KVIS = KINEMATIC VISCOSITY OF WATER
!K = COEFFICIENT OF SEDIMENT DISCHARGE EQUATION
!SHE = BED SHEAR STRESS
!CSHE = CRITICAL BED SHEAR STRESS
!F = VANONI'S (1977) PARAMETER TO DETERMINE CRITICAL SHEAR
!    STRESS
!A = DIMENSIONLESS SHIELDS PARAMETER OF CORRESPONDING F
!W = FALL VELOCITY OF SEDIMENT PARTICLES
!Z = FIRST TERM OF RUBEYS'S FALL VELOCITY FORMULA
!U = CONSTANT (=079) IN RUBEY'S FALL VELOCITY FORMULA
!B = SECOND TERM OF RUBEYS'S FALL VELOCITY FORMULA
!USTAR= SHEAR VELOCITY

!DECLERATION OF VARIABLES
    REAL K,SHE,USTAR,F,KVIS,A,CSHE
    REAL CUSTAR,U,B,Z,W
    DOUBLE PRECISION QSU
    K=12.5

    KVIS=0.000001

!CALCULATION OF BED SHEAR STRESS
    SHE=ROE*G*H*SEFF

!CALCULATION OS SHEAR VELOCITY DUE TO GRAIN ROUGHNESS
! ASSUMING HYDRAULIC RADIUS DUE TO GRAINS IS EQUAL TO FLOW DEPTH
```

```
      USTAR=(SHE/ROE)**0.5
!
! CALCULATION OF CRITICAL SHEAR STRESS

      F=(D50/(KVIS))*(0.1*(SS-1)*G*D50)**0.5

      IF(F.EQ.500.OR.F.GT.500) A=0.06
      IF(F.LT.500.AND.F.GT.400.OR.F.EQ.400) A=0.057
      IF(F.LT.400.AND.F.GT.300.OR.F.EQ.300) A=0.055
      IF(F.LT.300.AND.F.GT.200.OR.F.EQ.200) A=0.050
      IF(F.LT.200.AND.F.GT.150.OR.F.EQ.150) A=0.045
      IF(F.LT.150.AND.F.GT.100.OR.F.EQ.100) A=0.040
      IF(F.LT.100.AND.F.GT.90.OR.F.EQ.90) A=0.039
      IF(F.LT.90.AND.F.GT.80.OR.F.EQ.80) A=0.038
      IF(F.LT.80.AND.F.GT.70.OR.F.EQ.70) A=0.037
      IF(F.LT.70.AND.F.GT.60.OR.F.EQ.60) A=0.035
      IF(F.LT.60.AND.F.GT.50.OR.F.EQ.50) A=0.035
      IF(F.LT.50.AND.F.GT.40.OR.F.EQ.40) A=0.033
      IF(F.LT.40.AND.F.GT.30.OR.F.EQ.30) A=0.032
      IF(F.LT.30.AND.F.GT.20.OR.F.EQ.20) A=0.031
      IF(F.LT.20.AND.F.GT.15.OR.F.EQ.15) A=0.031
      IF(F.LT.15.AND.F.GT.10.OR.F.EQ.10) A=0.035
      IF(F.LT.10.AND.F.GT.9.OR.F.EQ.9) A=0.035
      IF(F.LT.9.AND.F.GT.8.OR.F.EQ.8) A=0.036
      IF(F.LT.8.AND.F.GT.7.OR.F.EQ.7) A=0.038
      IF(F.LT.7.AND.F.GT.6.OR.F.EQ.6) A=0.036
      IF(F.LT.6.AND.F.GT.5.OR.F.EQ.5) A=0.045
      IF(F.LT.5.AND.F.GT.4.OR.F.EQ.4) A=0.050
      IF(F.LT.4.AND.F.GT.3.OR.F.EQ.3) A=0.058
      IF(F.LT.3.AND.F.GT.2.OR.F.EQ.2) A=0.070
      IF(F.LT.2.AND.F.GT.1.OR.F.EQ.1) A=0.082

      CSHE=(SS*ROE*G-ROE*G)*D50*A


!
! CALCULATION OF CRITICAL SHEAR VELOCITY

      CUSTAR=(CSHE/ROE)**0.5

      IF(CUSTAR.GT.USTAR) THEN
      GOTO 1
      END IF
!
! CALCULATION OF FALL VELOCITY

      Z=(0.667+36*KVIS**2/(G*D50**3*(SS-1)))**0.5
      B=(36*KVIS**2/(G*D50**3*(SS-1)))**0.5

      IF (D50.EQ.0.001.OR.D50.GT.0.001) U=0.79
      IF (D50.LT.0.001) U=Z-B
      W=U*(D50*G*(SS-1))**0.5
!
! CALCULATION OF SEDIMENT DISCHARGE,QSU

      QSU=K*(ROS/(ROS-ROE))*SHE*(USTAR**2-CUSTAR**2)/W
      QSU=QSU/(ROE*G)
```

```
1    CONTINUE
     END

!32) COMPUTER PROGRAM CODE OF SUBROUTINE WUMOLIN

SUBROUTINE WUMOLIN(QU,H,G,D50,SS,QSU)

! KVIS= KINEMATIC VISCOSITY OF WATER
! W = FALL VELOCITY OF SEDIMENT PARTICLES
! A = FIRST TERM OF RUBEYS'S FALL VELOCITY FORMULA
! F = CONSTANT (=079) IN RUBEY'S FALL VELOCITY FORMULA
! B = SECOND TERM OF RUBEYS'S FALL VELOCITY FORMULA
! KSI = UNIVERSAL STREAM POWER
! C = SEDIMENT CONCENTRATION IN ppm
!

! DECLERATION OF VARIABLES

     REAL V,QU,H,A,SS,G,D50,KVIS,F,B,W
     DOUBLE PRECISION KSI,C,QSU

     V=QU/H

     KVIS=0.000001
! CALCULATION OF FALL VELOCITY

     A=(0.667+36*KVIS**2/(G*D50**3*(SS-1)))**0.5
     B=(36*KVIS**2/(G*D50**3*(SS-1)))**0.5

     IF (D50.EQ.0.001.OR.D50.GT.0.001) F=0.79
     IF (D50.LT.0.001) F=A-B
     W=F*(D50*G*(SS-1))**0.5
! CALCULATION OF KSI

     KSI=V**3/((SS-1)*G*H*W*(LOG10(H/D50))**2)

! CALCUATION OF SEDIMENT CONCENTRATION, C IN ppm

     C=1430*(0.86+KSI**0.5)*KSI**1.5/(0.016+KSI)

! CALCULATION OF SEDIMENT DISCHARGE, QSU

     QSU=C*QU/1000000

     RETURN
     END
```

Subject Index